高等职业教育教材

化工专业英语

冯艳文　李璐　主编

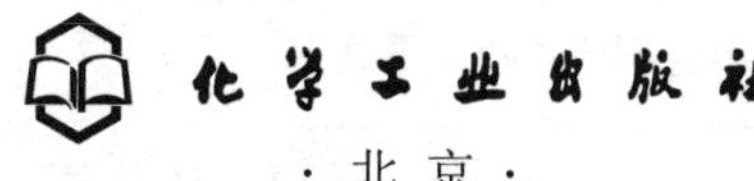

·北京·

内容简介

《化工专业英语》共六个项目，即化工专业英语概述、化学基础知识、化工基础知识、单元操作与控制、绿色化工与安全生产、专业英语应用与实践。本书以化工专业基础知识为载体，介绍化工专业常用的名词术语及常用句子的英文表达方式，同时针对性地介绍了化工专业英语的实际应用，内容丰富实用。本书选取的内容难度适中，具有一定的可读性，每课时配有关键单词、句子的音视频文件，可通过扫描书中的二维码获取。另外，部分章节有精选的习题并在书后附上答案，方便老师教学和学生自学。

本书可供高职高专和职教本科院校化工及其相关专业的师生学习使用。

图书在版编目（CIP）数据

化工专业英语 / 冯艳文，李璐主编. —北京：化学工业出版社，2022.9 （2024.8 重印）
ISBN 978-7-122-41465-6

Ⅰ.①化… Ⅱ.①冯…②李… Ⅲ.①化学-英语-高等职业教育-教材②化学工业-英语-高等职业教育-教材
Ⅳ.①O6②TQ-43

中国版本图书馆CIP数据核字（2022）第085914号

责任编辑：王海燕
文字编辑：曹　敏
责任校对：杜杏然
装帧设计：关　飞

出版发行：化学工业出版社
（北京市东城区青年湖南街13号　邮政编码100011）
印　　装：三河市双峰印刷装订有限公司
787mm×1092mm　1/16　印张 $15^1/_2$　字数380千字
2024年8月北京第1版第3次印刷

购书咨询：010-64518888　　售后服务：010-64518899
网　　址：http://www.cip.com.cn
凡购买本书，如有缺损质量问题，本社销售中心负责调换。

定　　价：45.00元

前言

化工专业英语是高等职业教育化工技术类专业开设的一门重要的专业必修课程。掌握和使用化工专业英语知识，从而具备化工专业英文科技信息获取和输出能力，是高等职业教育化工技术类专业培养复合型、发展型、创新型高素质技术技能人才必备的基础知识和技能之一。

编者在总结天津职业大学多年化工专业英语教学实践基础上，整合丰富的数字化资源，精心编写了《化工专业英语》教材。本教材将专业英语语法、词组的固定搭配、英文科技信息获取与输出实践等与高等职业教育化工技术类专业基础和核心技术技能学习相融合，设置了化工专业英语概述、化学基础知识、化工基础知识、单元操作与控制、绿色化工与安全生产以及专业英语应用与实践六个项目。采用项目化编排，以科技英文为载体，体现国内外新技术、新材料、新设备和新工艺，实现专业模块学习与英语运用能力提升的无缝融合，语言精练、图文并茂、实践性强。同时，本教材尝试挖掘行业英语教学内容中的德育因素和德育内涵，注重职业素养、创新能力、工匠精神、爱国情怀等思政教育元素的融入。

本教材实现“纸数”融合、多媒体共享，以纸质教材内容为核心，配套多样化的数字教学资源。纸质教材主体采用主、副栏版面设计，主栏作为主体部分，除理论知识的阐述，还包含有“长难句语法分析”“课后习题”“拓展阅读”等栏目，强化理论核心知识，详解语法重点难点，拓展专业产业前沿；副栏设置“重点单词词组”“链接二维码”等，且副栏排版与主栏中正文相匹配，方便学生在阅读的过程中同步学习词汇。数字教学资源方面，可运用二维码无缝链接音频、视频等媒体资源。注重英语教学实践性，突出对学生进行听力、会话、阅读、写作等训练。重点单词词组等不仅有实景例句，同时提供标准发音练习，使学生进行听、说、读全方位训练。

本书由天津职业大学冯艳文教授、李璐博士主编，天津职业大学尚亚平老师副主编，张发荣博士、张畅老师和韩琳老师参与编写。其中项目一、项目六由冯艳文、李璐和张畅编写，项目二由李璐编写，项目三、项目四由尚亚平编写，项目五由张发荣编写。全书的英文语法分析及英文校稿工作由张畅、韩琳负责。单词例句由李璐编写并配音。全书由冯艳文、李璐负责统稿，河北工业大学博士生导师张月成教授主审。天津职业大学胡兴兰教授为本书的策划和编写提供了大力支持。此外，在本书编写过程中得到了天津职业大学应用化工技术专业的领导、兄弟院校、化工企业的工程技术人员及化学工业出版社的大力支持，在此一并表示衷心感谢！

由于编者水平所限，书中疏漏在所难免，恳请读者批评指正。

编 者

2022 年 5 月于天津

目录

项目一
化工专业英语概述

Part Ⅰ:
Overview of Chemical Engineering Professional English (CEPE)

第一单元　化工专业英语特点
Unit 1　Characteristics of CEPE

In this lesson you will learn:

- Features of vocabulary
- Features of expression
- Features of grammar

随着化工行业的迅猛发展，中国化工企业正步入经济全球化、信息社会化、产业知识化的时代，国际贸易和学术交流逐渐加深并普遍化。我国企业正不断加强与境外企业的合作，实现国内外企业的协同效应，包括重构企业组织、交流企业的先进管理经验、加强双方人员交流与沟通、加速中外文化和管理的融合、加速新兴市场的开发等等。此外，相关研究显示，近八成的先进科学技术以及研究成果是以英文形式呈现的；除专业文献外，还包括专利文书、质检报告等各种应用文也涉及英文。同时很多生产及研发过程中使用的先进仪器和专业设备的相关技术说明的主要语言也均为英文。在国际贸易频繁的当今，化工专业英语在实际应用中发挥着越来越重要的作用。化工专业人才不仅要有扎实的专业知识，还要具备娴熟的专业英语交流能力和翻译能力。因此，学习化工专业英语对提高化工相关专业人才的基本素质和岗位工作能力具有重要的意义。

化工专业英语隶属于科技英语，是经过长期发展而形成的具有特殊用途的行业英语，化工专业英语使用具有专业性的英语语言来阐述化工专业理论知识、描述化学领域客观现象以及介绍化工专业学科发展等。这一体系具有其自身的语言特点，主要体现在词汇专业性强、文体结构严密和语法特征突出三个方面。本次课将从用词、表达和语法三个方面对化工专业英语的特点进行介绍。

一、用词特点

1. 专业词汇

在科技英语中，不同学科或专业都有各自的专业词汇或术语，以体现其专业性和科学性。化工专业英语更是如此，其词汇中存在着大量的与化学化工相关的专业词汇。例如：

Hydrogenation（加氢），liquefaction（液化），distillation（蒸馏）。

部分专业词汇意义单一明确，具有与之对应的中文表达，例如，aluminium（铝）、catalyst（催化剂）、ozone（臭氧）等。还有一些专业词汇在不同语境中具有不同的释义，例如，fraction 表示“部分，分数”等意义，但在化工领域 fraction 则表示分馏操作中的“馏分，组分”。可见，翻译时一定要结合具体语境，准确选择词义，避免说“外行话”。

2. 派生词

同时，化工专业英语词汇经常利用前缀、后缀构成派生词。例如化合物的系统命名法体系利用词汇的前缀、后缀对不同结构的化合物进行分类和命名。例如，methane 是甲烷，butane 是丁烷，前缀 meth-和but-分别代表了“一”和“四”; ethene 是乙烯，ethyne 是乙炔，这里后缀-ene和-yne 分别代表了烯烃和炔烃；又如，环戊酸是 cyclopropanecarboxylic acid，前缀 cyclo-代表环状结构，propane 代表环上有五个碳且均以单键相连，后半部分 carboxylic acid 则表示羧酸。

其他常见的派生词还包括-ate 常用作酸所形成的盐类或酯类的词根；-ide 通常表示化合物；前缀 de-表示“去除、脱除”等意思；bio-用作与生物相关联的专有名词的词根等。

3. 缩略词

此外，化工专业英语中还存在着很多缩略词，具有其特定的使用范围。例如，DCS（distributed control system）表示分布式控制系统；CCR（continuous catalytic reforming）表示连续催化重整；DCL（direct coal liquefaction）表示煤直接液化等。这些缩略词简洁明了，具有固定且准确的翻译，避免了冗长的词组，能够为读者提供准确的指向性。

二、表达特点

化工专业英语的表达形式与大学英语相比差异较大，其学习内容与化工专业知识理论密切相关。一般包括生产工艺过程、反应原理、实验现象、操作步骤、工艺条件以及分析方法等。这就决定了化工专业英语在表达形式上一定要符合专业客观性质的要求，具有较强的逻辑性，且结构严谨，行文简洁，注重清晰性、连贯性和流畅性，避免过多运用修辞比拟手法。

通常化工专业英语的表达具有句式长、结构复杂的特点。

1. 句式长

化工专业英语经常用到长句，常为无人称句式“it...”，涉及的专业术语较多，且经常出现化学反应式、结构图以及流程图等，需要学生具备一定的化工专业知识和相关英语词汇储备，才能很好地理解这类长句的专业意义。

例如：Turbulent flow is characterized by the large-scale, observable fluctuations in fluid and flow properties, and the shear stress is the result of these fluctuations. (*Fundamentals of Momentum*, *Heat and Mass Transfer* P85)

It is interesting to note that the constant internal energy assumption and the inviscid flow assumption must be equivalent, as the other assumptions were the same. (*Fundamentals of Momentum, Heat and Mass Transfer* P119)

It is observed that all objects emit electromagnetic radiation over a range of frequencies with an intensity that depends on the temperature of the object. (*Physical Chemistry* P713)

Objectionable hydrogen sulfide is removed from such a gas or from naturally occurring hydrocarbon gases by washing with various alkaline solutions in which it is absorbed.

2. 结构复杂

英语的语言体系重形合，句子逻辑关系严密，语义层次分明。化工类英文文献中经常出现关系严密的主从句以及排比句，以详细阐述某些复杂的客观现象、原理等。

例如：Because the standard reaction Gibbs energy is large and positive, you can anticipate that K will be small, and hence that $\alpha \ll 1$, which opens the way to making approximations to obtain its numerical value. (*Physical Chemistry* P644)

此外，有时英语句式结构、时间顺序、逻辑关系与汉语的表达方式、习惯等不完全一致甚至出现完全相反的情况，这些都导致了化工专业英语自身的多变性和复杂性。

三、语法特点

1. 语态

化工专业英语广泛使用第三人称的被动语态，这是由于专业领域关注的是专业原理、客观现象而非行为者；另外，被动语态句式将最重要的信息放在句首，比主动语态更直接明了。例如：Populations of animals can also be decimated by chemical effects that do not kill the animals. (*Introduction to Green Chemistry* P7)

Experimentally, changes in internal energy or enthalpy may be measured by techniques known collectively as “calorimetry”. (*Physical Chemistry* P164)

2. 时态

时态方面，化工专业英语的写作通常采用一般现在时，化工专业所涉及的研究背景、反应原理等一般没有特定的时间关系，使用一般现在时可以进一步突出专业上的客观性。

此外，过去时态也经常出现在化工专业英语中，用来描述已发生的客观事实，如某人某项目的研究方法、实验步骤、实验结论等。例如：He then tried to measure the change in temperature of the water of the bath when a stopcock was opened and the air expanded into a vacuum. (*Physical Chemistry* P242)

3. 非限定动词及名词化词组

为了使行文精练、结构紧凑，科技英语中大量使用非限定动词和名词化结构、名词和介词短语。动名词短语可用来替代时间、条件、伴随状语从句。例如：

分词短语可取代关系从句。例如：The molecules move apart from each other, like a ball rising from a planet.（现在分词短语代替关系从句，相当于：The molecules move apart from each other, like a ball that is rising from a planet.）(*Physical Chemistry* P258)

不定式短语可以替换目的状语从句。例如：To cool a gas, therefore, expansion must occur without allowing any energy to enter from outside as heat.（相当于：In order that a gas can be cooled, therefore, expansion must occur without allowing any energy to enter from outside as heat.）

名词化结构可以简化句式，从而使表述更加客观、信息更加确切。例如：We can normally regulate the temperature by using a thermo-couple.

可写为：Regulation of the temperature can normally be effected by the application of a thermo-couple. 这种写法将名词性结构和被动语态相结合，避免使用人称代词作主语，将重点更多地放在了客观现象上。

4. 后置定语

化工专业英语经常使用后置定语，包括介词短语、单个分词、定语从句等结构。例如：

In general, ethers are good solvents for fats, waxes and resins.（介词短语作后置定语）

The functional group of a ketone consists of a carbon atom connected by a double bond to an oxygen atom.（过去分词作后置定语）

A semiconductor is a material that is neither a good conductor nor a good insulator.（定语从句）

第二单元　化工专业英语常用语法

Unit 2　Grammar Frequently Used in CEPE

In this lesson you will learn:

- Five basic sentence patterns
- Clause
- Passive voice

一、化工专业英语五种句式

英语句子成分可以分为基本成分、附属成分、独立成分、省略成分和连接成分，其中最关键的是句子的基本成分，包括主语、谓语（不及物动词、及物动词、双宾动词、系动词、宾补动词）、宾语及补语，而我们熟悉的修饰语如定语、状语都不属于句子的基本成分。完整的英语句子一般至少包含 2 个基本成分，最多包含 4 个基本成分，组成的 5 个基本句型如下：

1. 主谓

即主语 + 不及物动词，这类句型中的谓语动词后不可以直接加宾语，句中只有主语和谓语两个主要成分。此句型的谓语动词都能表达完整的意思，后面还可以添加修饰语等其他成分。

例如：Electrolytes are constantly breaking down in water.

电解质在水中不断地分解。

2. 主谓宾

即主语 + 及物动词 + 宾语，这类句型中的谓语动词后面可以直接加宾语，也就是动作执行的对象。在主谓宾结构的基础上也可以添加其他修饰成分。

例如：Marie Curie discovered radium in 1898.

玛丽·居里于 1898 年发现了镭。

3. 主系表

即主语 + 系动词 + 表语。系动词也叫连系动词，用来连接主语和表示主语身份、性质、特征或状态的表语。表语通常由名词、形容词、介词短语、动名词、不定式来充当。在主系表结构的基础上也可以添加其他修饰成分。

例如：An acid is a compound whose solutions can produce hydrogen ions.

酸是这样一种化合物，其溶液能产生氢离子。

4. 主谓双宾

即主语 + 及物动词 + 间接宾语 + 直接宾语，其中直接宾语指动词直接作用的对象，而间接宾语是指动词需要先借助于一个间接的对象，再把动作传递到直接宾语身上。常见的带双宾语的及物动词有 give, ask, bring, offer, send, pay, show, tell, get 等。在主谓双宾结构的基础上也可以添加其他修饰成分。

例如：The professor gives her a chance to conduct the experiment.

教授给了她一个做实验的机会。

5. 主谓宾补

即主语 + 及物动词 + 宾语 + 宾语补足语，其中宾语补足语可以对宾语做出补充说明，表明宾语的身份、特征等，使句意完整，简称宾补。不定式、现在分词、过去分词、形容词、副词和介词短语等都可以作宾补。在主谓宾补结构的基础上也可以添加其他修饰成分。

例如：The oily material keeps the substance soft and pliable.

这种油性材料使物质保持柔韧。

二、从句

英语中主要有三大类从句，即名词性从句（包括主语从句、宾语从句、表语从句、同位语从句）、形容词性从句（即定语从句 / 关系从句）、副词性从句（即状语从句），这三类从句在句子中充当不同的成分，能够使一句话包含更多的信息量，提高表达效率，是化工英语中常见的语法结构。

1. 名词性从句

（1）主语从句　主语从句是在句中充当主语成分的从句，连接词包括：连接代词 what, who, whom, whose 等；连接副词 how, why, when, where 等；从属连词 that 和 whether。

例如：What is more important is the great increase in circuit reliability.

更为重要的是，电路的可靠性显著提高了。

值得注意的是，在化工专业英语中，主语从句很多情况下都可以放到句子后面，而用代词 it 作形式上的主语。

例如：It is believed at the present time that all so called magnetic phenomena result from forces between electric charges in motion.

现在人们确信，一切所谓磁现象都是由运动电荷之间的作用力引起的。

It is obvious that these techniques will be of great value in cryogenic switching circuit.

很显然，这种工艺对低温开关电路有很大的价值。

（2）宾语从句　宾语从句是在句中充当宾语成分的从句，当从句是陈述句时，从句用 that 引导；当从句是疑问句时，从句用 whether/if 引导；由特殊疑问词引导的宾语从句，特殊疑问词有 who, whom, whose, what, which, when, where, why, how 等。此外应特别注意从句的语序为陈述句语序。

例如：No one can say just what these basic positive and negative charges of electricity are.

还没有人能说明这些基本的正、负电荷究竟是什么。

Failure can be prevented by ensuring that all wiring is properly insulated.

防止这种事故的办法是保证让导线完全绝缘。

（3）表语从句　表语从句是在句中充当表语成分的从句，常见的引导词为 when, where, why, who, how, that, whether 等。

例如：Perhaps the most common classification of material is whether the material is metallic or non-metallic.

也许，材料最常见的分类是根据它是金属还是非金属来进行的。

One of the remarkable things about it is that the electromagnetic waves can move through great distances.

其显著特点之一就是电磁波能传播得很远。

（4）同位语从句　同位语从句是在句中充当同位语的从句，即重复说明同一件事物的从句，用来对前面的抽象名词进行解释说明。常见的连接词有 that, who, how, when, where 等。后接同位语从句的抽象名词通常有 news, idea, fact, promise, question, doubt, conclusion,thought, message, suggestion, possibility, decision 等。

例如：We all know the fact that all elements are made up of atoms.

我们都知道元素都是由同一种原子组成的这一事实。

The story how they invented this important device is quite interesting.

他们怎样发明这个重要装置的事，是十分有趣的。

2. 关系从句

关系从句，也被许多人称为定语从句，是由关系词引导的常做定语的从句，一般出现在它所修饰的词语之后，在化工专业英语中十分常见。关系从句中被修饰的词叫做先行词，在先行词后引导关系从句的关系词包括 that, which, who, whom, whose, as, than 等，关系词在从句中充当一定的成分。关系从句可分为两类：

（1）限制性关系从句　限制性关系从句起限定作用，修饰特定的人或事物，如果去掉限制性关系从句，整个句子的表意将会不完整，句子也会不通顺。从结构上来说，限制性关系从句一般紧跟先行词，与先行词之间往往没有逗号分隔。

例如：Mild oxidation of a primary alcohol gives an aldehyde which may be further oxidized to an organic acid.

伯醇轻度氧化生成醛，而醛能进一步氧化生成有机酸。

A conductor is a substance which is able to carry electrons easily.

导体是一种易于携带电子的物质。

（2）非限制性关系从句　非限制性关系从句主要起描写和补充说明的作用，通常含有原因、让步、时间、条件、结果等含义，从句要用逗号与主句分开，关系代词不能省略，通常用which（不用that），who, when等引导。

例如：Copper, which is used so widely for carrying electricity, offers very little resistance.

铜的电阻很小，因此被广泛用于输电。

Alloys, which contain a magnetic substance, generally also have a magnetic properties.

合金如果含有磁性物质，一般也会具有磁性。

3. 状语从句

状语从句在句中用作状语，起副词作用。状语从句可以修饰的对象非常广泛，包括谓语、非谓语动词、定语、状语和整个句子。其种类繁多，位置灵活，因此无论在日常英语还是在化工专业英语中，状语从句的应用都非常广泛，主要包含以下几种常见类型：

（1）时间状语从句　常见的连接词有when, after, before, until, as soon as, while, as, once等。

例如：The conductivity of this material increases as the temperature increases.

该材料的电导率随温度的上升而增加。

Nearly 100 years passed before the existence of subatomic particles was confirmed by experiment.

在过了将近100年后才由实验证明了亚原子粒子的存在。

（2）地点状语从句　常见的连接词包括where, wherever, everywhere等。

例如：Generally, air will be heavily polluted where there are factories.

一般来说，有工厂的地方空气污染就严重。

The special-purpose machine is used where factors like weight and power consumption are critical.

这种专用机器用在重量、功耗等因素要求高的地方。

（3）原因状语从句　常见的连接词包括because, for, as, since, now that, in that等。

例如：The equipment was not invented because it was needed in the war.

这种设备并不是因为战争的需要而发明的。

The positive charge of the nucleus is of greatest importance as it determines the chemical nature of atom.

原子核的正电荷是极为重要的，因为它确定了该原子的化学性质。

（4）让步状语从句　让步表示某一动作或状态与另一动作或状态在意义上有矛盾，但不妨碍事实的进行或实现。常见的连接词有although, though, however (no matter how), even if, even though, whether...or等。

例如：Small as they are, atoms are made up of still smaller units.

原子尽管很小，但都是由一些更小的单元组成的。

All substances, whether they are gaseous, liquid or solid, are made of atoms.

一切物质，不论是气体，液体还是固体，均由原子构成。

（5）方式状语从句 常见的连接词有 as, as if, as though 等。

例如：The molecules of a gas behave as though they were perfectly elastic bodies.

气体分子的性能好像它们是理想弹性体一样。

Electrons move round the nucleus as the earth moves round the sun.

电子绕原子核运动就像地球绕太阳运行一样。

（6）目的状语从句 常见的连接词有 in order that, so that, in case, lest, for fear that 等。

例如：The professor asked the students to hurry up with the reports in order that she could give remarks.

教授要求学生快写报告，这样她就能批阅了。

The magnet is usually made in the shape of horseshoes so that it will be as strong as possible.

磁铁通常做成马蹄形的，以便使磁性尽可能强。

（7）结果状语从句 常见的连接词有 so that, such that, so...that, such...that 等。

例如：The normal atom has an exactly equal number of positive and negative charges, so that the atom as a whole is neutral.

正常原子中正负电荷数是完全相等的，所以从整体上看原子是中性的。

The concept of element is so important that it will bear further discussion.

元素的概念极为重要，所以还需要进一步加以讨论。

（8）条件状语从句 常见的连接词有 if, unless, whether, as long as, only if, providing that, on condition that 等。

例如：No flow of water occurs through the pipe unless there is difference in pressure.

除非有压差，否则水是不会流过管子的。

So long as the units are the same on both sides of the equal sign, we have a valid equation.

只要等号两边单位相同，我们就得到了一个能成立的等式。

三、被动语态

语态是动词的一种形式，用来说明主语与谓语动词之间的关系。英语中有两种语态，主动语态和被动语态，其中被动语态是不知道动作执行者或强调动作承受者的一种语态。在被动语态中，主语是动作的承受者，即行为动作的对象。与主动语态相比，被动语态将所要说明的事物放在突出位置，能加强句子的客观色彩。同时，被动句用活动、作用、事实、现象等作主语，而主语一般位于句首，这样能立即引起读者对讨论对象的注意，简洁明了，在表达上效率更高。化工专业英语具有很强的专业性和客观性，通常强调普遍真理，描述过程、特性或功能，表达推理或假设等，着眼于逻辑推理的过程和演绎论证的结果，而不过分关注某一动作是谁发出的。因此，被动语态在化工专业英语中得到大量使用。

1. 被动语态的用法

被动语态的基本构成为“be+ 及物动词的过去分词（+by+ 动作执行者）”，可用于各种时态，时态的变化通过 be 动词的变化来实现，同时人称和数的变化也在 be 动词上体现，如：

The heat is transferred by circulating carbon dioxide gas.

热量由循环流动的二氧化碳气传送。

The chain reaction must be carefully controlled.

对链式反应必须谨慎控制。

Measures have been taken to diminish friction.

已经采取了一些措施来减少摩擦。

2. 使用被动语态的几种常见情况

（1）突出动作的对象：化工专业相关文献主要讨论的是动作的对象，而不是动作的执行者，有时使用主动语态会模糊句子的中心，分散读者的注意力，如“Electrolysis is now widely used”如果改为主动语态“Now we use electrolysis widely”，则主语 electrolysis（电解法）不够突出。

（2）化工专业相关文献中的定义、定理、原理等，往往没有动作的执行者，或不必说出动作的执行者，因此常常使用被动语态，如：The word “plastic” is used to describe something which is easily shaped.（译文：“塑料”一词被用来描述某种易于成型的东西。）

（3）用被动语态表述问题，语义更清晰，结构更简单，如：Electricity for powering a fuel-cell car is produced when electrons are stripped from hydrogen atoms at catalysis sites on the membrane surface.（译文：当电子从膜表面催化位置的氢原子上剥离时，燃料电池汽车的电力就产生了。）

第三单元　化工专业英语阅读与翻译

Unit 3　Reading and Translation of CEPE

In this lesson you will learn:

- Reading skills
- Translation of CEPE vocabulary
- Translation of sentences

一、化工专业英语阅读

化工专业英语的词汇难度较大，语法结构相对复杂，因此在阅读相关文章和文献材料时，应当掌握一定的阅读技巧，如精读、泛读、略读、寻读等。

1. 精读

精读即逐字逐句阅读，通过对较短篇幅文章的深入分析和梳理，精确理解文中的词汇、语法、习惯表达和英文思维模式。在精读中，阅读者应详细深入地了解文章的全部内容，特别是细节性内容，并选出重点词句进行分析，记录下自己认为重要、有用的词汇和句式。对于化工专业英语学习者来说，在精读中尤其要注意积累化工专业词汇和术语，掌握化工专业英语中常见的句法结构和语法现象。

2. 泛读

泛读是一种迅速、广泛接受文字信息的阅读方法，通过大量浏览英文材料，经过日积月累使阅读者自然习得语言能力。泛读不要求学习者精研细读，也不需要记忆或背诵，只需要了解读物的性质和大致内容即可。与精读相比，泛读对阅读量的要求更高，需要学习者广泛涉猎领域内的各类文字材料，让反复出现的词汇和习惯表达自然而然地在头脑中生根发芽。化工专业英语学习者在泛读训练时应注意选择适当的材料，在训练初期阅读难度适宜、自己感兴趣的读物，从简易的科普性质读物逐渐过渡到专业性强的文献资料。

3. 略读

略读指的是快速浏览全文掌握主旨的阅读方法，是快速阅读中常用的技巧。在查阅文献资料时，读者有时只需要了解文章的主题思想，然后再对文章进行筛选和分类，这时就可以使用略读法，对文章的标题、首段、尾段、每段的段首句和结尾句进行略读，一般来说，文章内容的概括性陈述会出现在这些位置。略读可帮助读者快速了解文章主题，对文章的结构获得一个整体概念，对各部分内容得到一个粗略印象，使读者对文章主旨做出判断，方便读者对某些重要细节信息所在的段落进行快速定位。

4. 寻读

寻读是指以问题为线索，带着问题去寻找某一特定信息的阅读方法，也是快速阅读的常用技巧之一。寻读在日常生活中随处可见，比如查找特定的高铁车次、时间，在杂志中查找特定的栏目，在字典中查找词汇，在购物平台寻找自己心仪的商品，通过互联网查找招聘信息等都属于寻读的范畴。寻读需要读者在众多的资料信息中迅速查找和锁定目标信息，常见的定位词或线索词包括数字、年代、第一次出现的人名地名等专有名词、表明事物特征或比较关系的词等。

二、词汇的翻译技巧

化工专业英语的词汇专业性较强，使用范围特定，其词意相对单一明确。一般采用音译法、意译法、缩略法等进行翻译。

1. 音译法

音译法是指翻译人员根据单词的英文发音来对生僻的词汇进行与发音相似的翻译。随着科学技术的不断发展进步，对于专业性较强的化工领域而言，其专业术语也在不断拓展，音译法成为目前翻译过程中最为常见的翻译手段之一，被广泛应用于翻译工作中。例如化学音译用字“哌”可用来翻译英文中“piper”开头的化学物质，多代表含氮的饱和六元环，如哌啶（piperidine）和哌嗪（piperazine）。

2. 意译法

意译法是指在对原有语言理解的基础上，将全新的专业词汇进行转换和意译。作为最常见和普遍的翻译手段，意译法不仅能够使生涩的专业词汇和技术语言翻译更加清晰和精准，且有利于新词汇的记忆和理解。

对于表达意义单一明确，且有相对应的中文表达的词汇，采用直译的原则翻译即可。例如“hydrogen”对应化学元素氢；但是有一些词汇在不同的语境中有不同的表达意义，比如“element”一般译作“要素”“成分”，但在化工英语中译作“元素”，又如“carrier”通常译作“搬运工人/车/船”等，而在化工英语中译作“载体”。对于这类词汇，则要求翻译人员要明确地掌握并理解其在化工专业中的概念，避免造成不恰当的翻译。

3. 缩略法

在化工专业文献中，经常会出现一些英语词汇，其全称复杂且冗长，不利于记忆和书写。采用缩写的方式代替较为复杂的词汇，不仅有利于记忆，且使用和书写更加快速便捷，有利于翻译的工作顺利进行。例如：BAA（正丁醛苯胺缩合物）、PAPI（多亚甲基多苯基多异氰酸酯）、PVC（聚氯乙烯）等。

应当指出的是，许多缩略词的英文形式虽然相同，但其含义在不同的专业领域却大相径庭。例如：BAA 也可以指英国会计协会（British Accounting Association），PAPI 也指精密进近航道指示器（Precision Approach Path Indicator），PVC 也指心室早期收缩（Premature Ventricular Contraction）。对于这样的缩略词，翻译时要特别注意结合具体语境，选择正确的词义，避免不必要的误解。

三、句子的翻译技巧

1. 被动句的翻译

由于化工专业英语的严谨客观性等特点，被动语态在文献中的使用频率很高，对于这种句式的理解和翻译非常重要。

被动句的翻译主要有以下方式，第一是把被动句译为主动句，例如：If one or more electrons are removed, the atom is said to be positively charged.（译文：如果原子失去了一个或多个电子，我们就说这个原子带正电荷。）译文中将被动语态改为主动语态，更符合汉语的表达方式，也更容易被读者接受。第二种方式是译为汉语被动句，通过“被、由、让、遭、为……所、加以”等词来表达英语被动语态中的助词含义。例如：The atomic theory was not accepted until the last century.（译文：原子学说直到上个世纪才为人们所接受。）此外，也可以根据语境，翻译为汉语的无主句，将英语句中的主语译成汉语中的宾语，即先译谓语，后译主语。例如：Much greater magnification can be obtained with the electron microscope.（译文：使用电子显微镜，能获得更大的放大倍数。）

2. 长难句的翻译

在化工专业英语的翻译中，长难句的翻译一直都是难点所在。中文和英文两种语言体系自身存在巨大的差异，汉语重意合，结构上较为松散，英语重形合，结构更紧凑，逻辑思维较为严谨。处理专业英语中的长难句时，应充分考虑到中西方思维的差异，根据语境、句法内容与特点等，灵活选用翻译策略，按照目的语的表达习惯、语法规则将原文准确、流畅地表达出来。

翻译英语长难句时，根据原文的特点可采用顺译法、倒译法和分译法三种方式进行翻译。

顺译法：英文句子的句法结构、逻辑关系、时间顺序等与中文表达习惯基本一致时，可直接按照原文的语序进行翻译。例如：The continuous process although requiring more carefully designed equipment than the batch process, can ordinarily be handled in less space, fits in with other continuous steps more smoothly, and can be conducted at any prevailing pressure without release to atmospheric pressure.（译文：虽然连续过程比间歇过程要求更为周密设计的设备，但连续过程通常能节约操作空间，较顺利地适应其他连续操作步

骤，并能在任何常用的压力下进行，而不必暴露在大气中。）例句中的英文的语序与译文的语序一致，每一部分的位置都可以一一对应，属于典型的顺译法。

倒译法：倒译法是指部分或完全改变语序，使译文更符合中文表达习惯。例如：This is why the hot water system in a furnace will operate without the use of a water pump, if the pipes are arranged so that the hottest water rises while the coldest water runs down again to the furnace.（译文：如果把管子装成这个样子，使最热的水上升，而最冷的水再往下回流到锅炉里去，那么，锅炉中的热水系统不用水泵就能循环，道理就在于此。）译文将 if 引导的从句放在句首，而把原本放在句首表示原因、理由的“why”放在句尾，调整了语序，读起来更通顺。

分译法：化工专业英语的长句中往往包含多个从句、不定式短语和分词短语等，如按照原文翻译成一句话，会显得整个句子非常复杂。这时可采用分译法，将原句拆分为多个短句，使译文更加清晰易懂。例如：Fire is a chemical reaction in which atoms of the gas oxygen combine with atoms of certain other elements, such as hydrogen or carbon.（译文：燃烧是一种化学反应，在这个反应中，气体氧的原子与某些其他元素的原子，如氧或碳，相化合。）例句中的句子成分复杂，有介词短语和定语从句，其中，which 引导的定语从句较长，为适应汉语的表达习惯，将其拆分为短句。

在化工专业英语的实际翻译当中，往往是顺译、倒译和分译互相融合，翻译者要对不同的翻译技巧融会贯通，根据文体和句式特点将翻译技巧灵活自然地运用于复杂的实际翻译中。

第四单元　化工专业英语写作与会话

Unit 4　Writing and Speaking of CEPE

In this lesson you will learn:

- Features of CEPE writing
- Common sentence patterns
- Features of CEPE speaking
- CEPE conversion example

一、化工专业英语的写作要点

在前面的章节中，从表达、语法、用词等方面论述了化工专业英语的特点，并讲解了英译汉的一些技巧。本节我们将以化工领域的两篇论文摘要为例，简要讲解一下化工专业英语应用文写作的特点、步骤及结构要求。

例一：

Phenolic resins are usually selected as carbon precursor because they give high production yield and allow the production of porous carbon with high surface areas and tailored porosity. With the increasing concern on fossil fuel depletion and environmental footprint, there is a strong global interest to explore renewable alternative of phenol-formaldehyde resins. In this work, alkaline poplar bark extractive-based phenolic resins were prepared. Using this bio-based phenolic resin as carbon precursor, a hierarchical porous carbon material with mciro/meso/macro porosity was obtained. Due to its unique hierarchical pore texture, the prepared carbon shows superior capacitive performance with high specific capacitance, excellent rate capability and good long-term cycle stability in KOH electrolyte. We believe that this bio-based activated carbon is a kind of economical and promising material because of the simple preparation process, low cost, environmentally friendly and high performance.

例二：

A hierarchically structured nitrogen-doped porous carbon is prepared from a nitrogen-containing isoreticular metal-organic framework (IRMOF-3) using a self-sacrificial templating method. IRMOF-3 itself provides the carbon and nitrogen content as well as the porous structure. For high carbonization temperature (950 ℃), the carbonized IRMOF-3 required no further purification steps, thus eliminating the need for solvents or acid. Nitrogen content and surface area are easily controlled by the carbonization temperature. The nitrogen content decreases from 7 at% to 3.3 at% as carbonization temperature increases from 600 to

950 ℃. There is a distinct trade-off among nitrogen content, porosity, and defects in the carbon structure. Carbonized IRMOFs are evaluated as supercapacitor electrodes. For a carbonization temperature of 950 ℃, the nitrogen-doped porous carbon has an exceptionally high capacitance of 239 F/g. In comparison, an analogous nitrogen-free carbon bears a low capacitance of 24 F/g, demonstrating the importance of nitrogen dopants in the charge storage process. The route is scalable in which multi-gram quantities of nitrogen-doped porous carbons are easily produced.

1. 写作特点

在用词方面，这两段文字使用了一定数量的专业词汇，且部分专业词汇来自拉丁语或希腊语，例如：phenolic, micro, meso, macro 等，这些词具有准确的意义，使用范围特定。另外，这段文字还含有添加前缀或后缀而成的派生词。例如：bio-base, long-term, nitrogen-doped, metal-organic, self-sacrificial, multi-gram，electrolyte, footprint, isoreticular, supercapacitor 等。第三个特点是复合名词的使用，例如：surface areas, production yield, fossil fuel, carbon structure 等，这些复合名词使得文字紧凑、干净、利落。

语法方面，大量使用一般现在时表述自然现象、过程、常规等“无时间性”的“一般叙述”。例如：Phenolic resins are usually selected as carbon precursor because they give high production yield and allow the production of porous carbon with high surface areas and tailored porosity. 对于项目组已做的实验的描述则使用一般过去式，例如：In this work, we prepared alkaline poplar bark extractive-based phenolic resins. 而且，文中多使用动词的被动语态，使论述显得更加客观，便于使读者的注意力集中在主体对象上。例如：A hierarchically structured nitrogen-doped porous carbon is prepared from a nitrogen-containing isoreticular metal-organic framework (IRMOF-3) using a self-sacrificial templating method; nitrogen content and surface area are easily controlled by the carbonization temperature. 此外，广泛使用不定式、动名词和分词等非限定性动词也是这两段文字的特点。例如，With the increasing concern on fossil fuel depletion and environmental footprint, there is a strong global interest to explore renewable alternative of phenol formaldehyde resins. 这些表达方式有助于行文在缺少各种修辞格的情况下保障文章的渲染力，增加其逻辑性和严密性。

句法方面，这些文字中经常出现同位语、定语从句、主语从句、宾语从句等从句结构。例如：In comparison, an analogous nitrogen-free carbon bears a low capacitance of 24 F/g, demonstrating the importance of nitrogen dopants in the charge storage process; the nitrogen content decreases from 7 at% to 3.3 at% as carbonization temperature increases from 600 to 950 ℃. 此外，文中经常使用名词化结构来强调存在的事实，而非某一行为。例如：For high carbonization temperature (950 ℃), the carbonized IRMOF-3 required no further purification steps, thus eliminating the need for solvents or acid. 这样的句法结构行文简洁、表达客观、确切严谨。

篇章结构方面，两段文字都体现了组织严谨、逻辑性强、层次清楚的特点。

2. 摘要的写作技巧

摘要往往是作者在一篇论文其他部分完成后写的，其内容分别从前言、实验方法、结构与讨论以及结论中选取关键内容进行概述。摘要的阅读好比给人第一口苹果的品尝效果，直接决定了读者是否继续阅读该篇论文的其他部分。此外，摘要中包含关键词，对论文的检索非常重要，且摘要的读者面比论文全文的读者面大得多。

（1）论文的主体结构；适当强调研究中的创新、重要之处，但不要使用评价性语言和带有感情色彩的语言；尽量包括论文的主要论点和重要的论证、数据等。

（2）使用简短的句子，用词应专业并为潜在的读者所熟悉；注意表述的逻辑性，尽量使用指示性的词语来表达论文的不同部分。

如使用“研究表明……”（It was found that...）表示结果；使用“通过对……的分析，认为……”（Based on..., it is suggested that...）表示讨论结果的含义等。

（3）确保摘要的“独立性”，避免引用文献、图表和缩写等；如果无法回避使用引文，应在引文出现位置将引文的书目信息标注在方括号内。

（4）避免在行文中穿插使用化学结构式、数学表达式、特殊符号等，以便于检索系统检索。

（5）摘要中的时态、语态应与在论文各部分中的时态语态相一致。一般采用现在时或过去时。

（6）不要使用第一人称。

二、英语论文写作常用词汇和句式

化工专业英语论文的写作基本包括摘要（abstract）、前言（introduction）、实验方法（experimental）、结果与讨论（result and discussion）以及结论（conclusion）几部分。要写好这一类型的文章，需要建立一个适合自己需求的句型库。本节以英文期刊发表的论文为例，介绍英文论文常用的词汇和句式。

1. 摘要部分

（1）回顾研究背景，常用词汇有 review, summarize, present, outline, describe 等。

（2）论文写作目的，常用词汇有 purpose, attempt, aim 等，另外还可以用动词不定式充当目的状语。

（3）介绍论文的重点内容或研究范围，常用词汇有 study, present, include,

focus, emphasize, emphasis, attention 等。

2. 前言部分

前言一般有两个作用，即介绍一个研究领域的发展状况和引起读者的兴趣。时态方面，一般采用现在时态，引用参考文献的句子可以使用过去时态。

（1）介绍前期工作，常使用“have been investigated”“have shown that”“It has been proved that...”“According to...”以及“...found/claimed...”等。

（2）阐述前期工作的局限性，可使用“However, little information/ work/ attention/ investigation has been...”“Even though considerable efforts have been devoted to..., rather less attention has been paid to...”“Therefore, further investigations are needed in order to...”等。

（3）介绍研究目的、研究意义和主要成果的常用句型包括：“The aim of this paper is to...”“The focus of the present paper is to...”“In this paper we report the results on...”“...sheds light on...”“...demonstrated an effective way to...”“...is discussed/ developed/ obtained...”“To gain an insight of...”等。

3. 结果与讨论部分

一般情况下，实验结果与讨论结合在一起进行阐述。

（1）描述实验结果时常用的连接词包括 therefore, hence, however, nevertheless, for instance, for example, furthermore, in addition, in comparison, in other words, similarly, alternatively 等。

（2）对于重要数据的肯定性的阐述和强调，可采用：“It is shown that...”“...is due to the fact that...”“The result proves/confirms/validates/...”“...is supported by...”“...has significant influence on...”等。

（3）表达不确定性的句型包括：“caused by...”“attributed to...”“contributed to...”“The data suggested/indicated/assumed...”等。此外，还可以加上不确定性修饰词，如“probably/could/may/appear to/tend to/seem to/might/apparently”等。

（4）对于异常实验结果的解释可采用以下表述：“The unexpected results could be due to...”“The difference may be explained as...”“The anomaly in the observations seems to come from...”“The errors may be due to the fact that...”。

（5）关于图表的描述：“Table 1/ Figure 1 shows/gives/provides/describes...”“...is shown/given/provided/stated in Table 1/Figure 1”“As shown in Table 1/Figure 1...”。另外，对于曲线图趋势的阐述，常用以下词汇：“increase/decrease/rise/decline/descend/ascend/fall off/steep fall/sharp rise/steadily climb/fluctuate”等。

4. 结论部分

结论部分是在论文的最后对研究内容、价值和重要性的总结，应注意与前言提及的研究意义相呼应。通常通过对重要数据的描述和概括性阐述，来展示所做研究的创新点，并为今后的研究指明方向。常用的句式包括：“In conclusion...”“In summary...”“...has demonstrated/shown for the first time that...”“...has provided the new perspective...”“The results lead to...”“...extends previous work on...”。指出未来研究方向的句式包括：“To further improve..., ...should be tested/studied/investigated”“To further verify..., ...is required”“Future research should be attempted to...”等。

三、化工专业英语口语会话特点

英语口语包括日常会话、即兴发言、独白、讨论和演讲等，其中应用于口头交际的日常会话与正式的书面语体相比，在使用上更加随意自由，在词汇和语法结构等方面具有鲜明的特点。

从词汇角度来看，英语口语会话使用的词汇量比书面语少，具有简洁、生动、通俗易懂的特点，其中还包含一些口语色彩很强的词汇，如缩略语、模糊语、语气词、时髦语、歇后语、俚语、方言、短语动词等。以下是几类具有代表性的口语词汇：

1. 用浅显、简短的单词代替较长的单词

obtain → get　　content → get　　genuine → real

attempt → try　　assistance → help

2. 用短语动词代替动词

return → go back　　postpone → put off

investigate → look into　　refuse → turn down

3. 缩略语

going to → gonna　　want to → wanna

got to → gotta　　kind of → kinda

从语法结构的角度来看，英语口语会话使用的结构相对简单，语法比较随意，说话人可以随时更换用词、句型，修正错误，填补句子，甚至使用不合乎语法规范的句子。多人会话中还常出现犹豫、停顿、插话等情况。英语口语会话多短句，少长句，句式结构多为简单句和结构简单的复合句，而结构复杂的复合句则相对较少，比如“He can’t go because he is busy.”在口语中可用两个简单句表达为“He can’t go. He’s busy.”同时，在较长的句子中，口语会话很少使用分词结构作状语，而更多选用并列句或从句，比如“Seeing the look on his face, she began to cry.”在口语中可表达为“She began to cry when she saw the look on his face.”

在英语日常会话的基础上，化工专业英语口语会话还具有专业性强的特点，需要说话者注意积累化工专业词汇、术语和习语。建议化工专业英语学习者使用国外相关真实语料加以模仿，在语境中理解和记忆。除此之外，在实际交流中，由于化工专业英语口语难度较高，为达到沟通目的，说话者应利用口语灵活变通的特点，采用表情、手势、动作、翻译工具甚至书写等方法促进理解，不必为个别词句钻牛角尖。

M4-1
校园场景会话

四、化工专业英语场景会话示例

A conversation between two students in a chemistry class：

Marie：Huh! This isn't working.

Peter：What isn't?

M：This reaction. I'm supposed to get hydrogen gas, but I don't seem to be getting anything except air.

P：How are you supposed to tell?

M：Well, hydrogen gas should explode when I hold this burning splint up to the test tube—but... nothing happens, see?

P：Hmm. Very unimpressive. How do you do this experiment, again?

M：Well, I put some hydrochloric acid into this test tube here. And then I add a piece of zinc metal to it. A reaction takes place and hydrogen gas is produced—it says—which should come out of the glass tubing, over here, and go into this other test tube. But it doesn't.

P：Nothing happens?

M：Well, something happens. The zinc bubbles and bounces around—the acid does something to it. Some gas must be coming off, I guess, but it isn't flammable. It doesn't burn at all.

P：Is there any way else to identify hydrogen?

M：Not that I've learned. It's completely colorless and tasteless, I know—it just burns very dramatically.

P：Huh. Well, I dunno. Shall we run through your experiment again together and see how it goes?

M：Sure, if you don't mind—that would be great! I'm sure I'm doing everything right, I've gone through it twice now, but I still get nothing.

P：OK. Let's see.... we need what? Clean test tubes? Two of 'em? Here we are. And... a rubber stopper with ten centimeters or so of glass tubing through it... right...uh...here. That's it?

M：Yep. And a graduated cylinder and the Bunsen burner. And the zinc and the acid.

P：All right. Let's do this first—let's take the zinc and the HCl from different sources—from, uh, those reagent bottles on that lab table over there. Maybe your bottles are contaminated or mislabeled or something.

M：Oh, good idea. Trust me to use the wrong reagents.

P：Just a sec.... here we are... OK, now you run the experiment like you think you should, and I'll watch what you do.

M：Well, first I clamp this test tube onto its stand...like this... and put in about five milliliters of the hydrochloric acid....

P：“About” five milliliters?! Isn’t this a scientific laboratory?

M：All right, all right. Exactly five milliliters of HCl...like this....

P：OK. Mmm... and....

M：Then I take a little piece of zinc metal with these tweezers, like this, and—

P：Are they all the same size?

M：Yeah, they seem to be. It looks like they’re pressed out of some kind of machine in standard bits.

P：And add “one”?

M：Yeah, one. Like this. And see?—it starts to bubble!

P：Sure does!

M：So I slap the stopper into the top of the test tube real quick, like—unh!—this....

P：Mmm....

M：And then, I hold the other test tube under the other end of the glass tubing, here...like this, and...uh, wait until I think I’ve collected enough hydrogen gas to ignite, I guess. Is the Bunsen burner going?

P：Yes, It’s burning—and you also need a splint, right? Here’s one. Hey! Just a minute—how in heck are you going to collect any gas like that?!

M：Like what?

P：Like that—holding the test tube under the mouth of the tubing.

M：What? Am I spilling it or something?

P：No, no—you’re losing it entirely! Hydrogen is lighter than air, Marie. It’s floating up.

M：Huh?

P：It’s floating up. It’s rising from the tube mouth. It’s not falling into your container. Hold it above the tube!

M：Ack! How stupid! I am so stupid. I’m never gonna pass this course. OK, OK—now...light that for me, will you?

P：Right...here you go.

M：And.... POW! Wow, did you see that?

P：And we’ve got lift off! And you’ve got hydrogen. Congratulations.

M：Yeah. (sighs) Thanks, Peter. Now all I gotta do is write this experiment up.

P：No problem. So, what’s your next experiment?

M：Uh, something about sodium and water.

P：Oh no!

项目二 化学基础知识

Part Ⅱ: Fundamental Knowledge of Chemistry

第五单元　化学元素与物质

Unit 5　Chemical Elements and Substances

Lesson one: Elements and Periodic Table

In this lesson you will learn:

- Introduction of element
- Periodic table
- Periodic law of the element

1. The elements

substance　*n.* 物质，材料

distinctive　*adj.* 有特色的，与众不同的

abbreviate　*vt.* 使缩短，缩减

M5-1
单词读音及例句

The term element refers to a pure substance with atoms, all of a single kind. To the chemist the "kind" of atom is specified by its atomic number, *Z*, since this is the property that determines its chemical behavior. Each chemical element has been given a name and a distinctive symbol as shown in Figure 5-1.

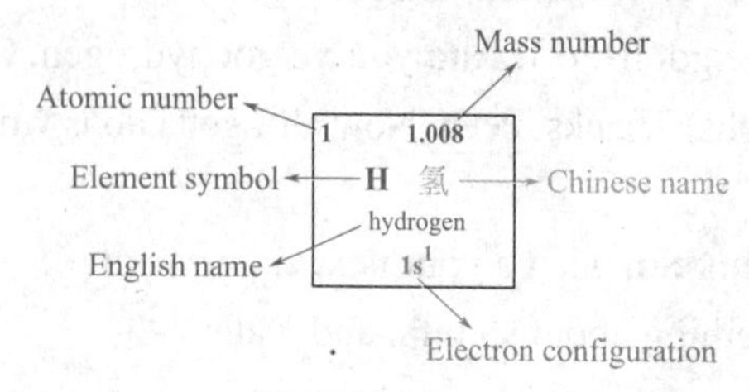

Figure 5-1　The chemical element symbol

At the end of 2016, the number of elements in the periodic table stood at 118. At the end of 2015, the IUPAC verified the discovery of elements with atomic numbers 113, 115, 117 and 118.

For most elements the symbol is simply the abbreviated form of the English name consisting of one or two letters, for example:

oxygen=O　nitrogen=N　neon=Ne　magnesium=Mg

Some elements, which have been known for a long time, have symbols based on their Latin names, for example:

iron=Fe (ferrum)　copper=Cu (cuprum)　lead=Pb (plumbum)

小提示：

IUPAC—The International Union of Pure and Applied Chemistry

国际纯粹与应用化学联合会

To find out more: https://iupac.org/

2. The periodic table

Beginning in the late 17th century with the work of Robert Boyle, who proposed the presently accepted concept of an element, numerous investigations produced a considerable knowledge of the properties of elements and their compounds. In 1869, D. Mendeleyeev and L. Meyer, working independently, proposed the periodic law. In modern form, the law states that the properties of the elements are periodic functions of their atomic numbers.

In other words, when the elements are listed in order of increasing atomic number, elements having closely similar properties will fall at definite intervals along the list. Thus it is possible to arrange the list of elements in tabular form with elements having similar properties placed in vertical columns. Such an arrangement is called a period as shown in Figure 5-2.

s-block elements		d-block elements										p-block elements					
Group 1	Group 2	Group 3	Group 4	Group 5	Group 6	Group 7	Group 8	Group 9	Group 10	Group 11	Group 12	Group 13	Group 14	Group 15	Group 16	Group 17	Group 18
1 H																	2 He
3 Li	4 Be											5 B	6 C	7 N	8 O	9 F	10 Ne
11 Na	12 Mg											13 Al	14 Si	15 P	16 S	17 Cl	18 Ar
19 K	20 Ca	21 Sc	22 Ti	23 V	24 Cr	25 Mn	26 Fe	27 Co	28 Ni	29 Cu	30 Zn	31 Ga	32 Ge	33 As	34 Se	35 Br	36 Kr
37 Rb	38 Sr	39 Y	40 Zr	41 Nb	42 Mo	43 Tc	44 Ru	45 Rh	46 Pd	47 Ag	48 Cd	49 In	50 Sn	51 Sb	52 Te	53 I	54 Xe
55 Cs	56 Ba	57-71 La-Lu	72 Hf	73 Ta	74 W	75 Re	76 Os	77 Ir	78 Pt	79 Au	80 Hg	81 Tl	82 Pb	83 Bi	84 Po	85 At	86 Rn
87 Fr	88 Ra	89-103 Ac-Lr	104 Rf	105 Db	106 Sg	107 Bh	108 Hs	109 Mt	110 Ds	111 Rg	112 Cn	113 Nh	114 Fl	115 Mc	116 Lv	117 Ts	118 Og

f-block elements

Lanthanoids	58 Ce	59 Pr	60 Nd	61 Pm	62 Sm	63 Eu	64 Gd	65 Tb	66 Dy	67 Ho	68 Er	69 Tm	70 Yb	71 Lu
Actinoids	90 Th	91 Pa	92 U	93 Np	94 Pu	95 Am	96 Cm	97 Bk	98 Cf	99 Es	100 Fm	101 Md	102 No	103 Lr

Figure 5-2　The periodic table

Notes:

When the elements are listed in order of increasing atomic number, elements having closely similar properties will fall at definite intervals along the list.

译文：当元素按原子序数递增的顺序排列时，性质极为相似的元素将按一定的间隔排列。

语法：句中 when 为连词，引导时间状语从句；现在分词短语 having closely similar properties 做名词 elements 的后置定语。

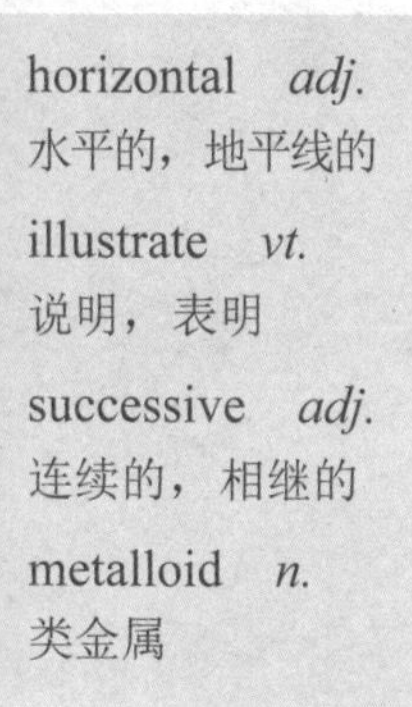

horizontal *adj.* 水平的，地平线的

illustrate *vt.* 说明，表明

successive *adj.* 连续的，相继的

metalloid *n.* 类金属

M5-2 单词读音及例句

transition metal 过渡金属

characteristic *n.& adj.* 特征，典型的

M5-3 单词读音及例句

Each horizontal row of elements constitutes a period. It should be noted that the lengths of the periods vary. There is a very short period containing only 2 elements, followed by two short periods of 8 elements each, and then two long periods of 18 elements each. The last two periods are long periods and both of them contain 32 elements.

With this arrangement, elements in the same vertical column have similar characteristics. These columns constitute the chemical families or groups. The groups headed by the members of the two 8-element periods are designated as main group elements, and the members of the other groups are called transition or inner transition elements.

In the periodic table, a heavy stepped line divides the elements into metals and nonmetals. Elements to the left of this line (with the exception of hydrogen) are metals, while those to the right are nonmetals. This division is for convenience only; elements bordering the line, the metalloids, have properties characteristic of both metals and nonmetals. It may be seen that most of the elements, including all the transition and inner transition elements, are metals.

The periodicity of chemical behavior is illustrated by the fact that, excluding the first period, each period begins with a very reactive metal. Successive elements along the period show decreasing metallic character, eventually becoming nonmetals, and finally, in group ⅦA, a very reactive nonmetal is found. Each period ends with a member of the noble gas family.

Reading comprehension

1. What property determines the chemical behavior of the elements?

2. How many elements does the shortest period contain?

3. How does the periodic table arrange the elements which have similar characteristics?

4. Write the distinctive symbols for elements of the noble gas family.

Exercise

1. True or false.

（1）Some elements have symbols based on their Latin names.

（2）Each period begins with a very reactive metal.

（3）Some elements have properties characteristic of both metals and nonmetals.

2. Translate the following sentences into English or Chinese.

（1）With this arrangement, elements in the same vertical column have similar characteristics.________________________________

____

____

（2）大多数元素，包括所有的过渡元素，都是金属元素。________

____

小提示：
- with... 引导方式状语从句，译作“通过……”
- including 包括……

Further Reading

The Periodicity of Elements

Except for hydrogen, a gas, the elements of group ⅠA make up the alkali metal family. They are very reactive metals, and they are never found in the elemental state in nature. However, their compounds are widespread. All the members of the alkali metal family form ions having a charge of +1 only. In contrast, the elements of group ⅠB—copper, silver, and gold—are comparatively inert. They are similar to the alkali metals in that they exist as 1+ ions in many of their compounds. However, as is characteristic of most transition elements, they form ions having other charges as well.

The elements of group ⅡA are known as the alkaline earth metals. Their characteristic ionic charge is +2. These metals, particularly the last two members of the group, are almost as reactive as the alkali metals. The group ⅡB elements—zinc, cadmium, and mercury are less reactive than those of group ⅡA, but are more reactive than the neighboring elements of group ⅠB. The characteristic charge on their ions is also +2.

With the exception of boron, group ⅢA elements are also fairly reactive metals. Aluminum appears to be inert toward reaction with air, but this behavior stems from the fact that the metal forms a thin, invisible film of aluminum oxide on the surface, which protects the bulk of the metal from further oxidation. The metals of group ⅢA form ions of +3

alkali metal 碱金属

alkaline earth metal 碱土金属

widespread *adj.* 分布广的，普遍的

M5-4
单词读音及例句

in contrast 相比之下

ionic charge 离子电荷

stem from 来自，起源于

M5-5
单词读音及例句

nonelectrolyte *n.* 非电解质

nitride *n.* 氮化物

phosphide *n.* 磷化物

arsenide *n.* 砷化物

halogen *n.* 卤素

M5-6
单词读音及例句

charge. Group ⅢB consists of the metals scandium, yttrium, lanthanum, and actinium.

Group ⅣA consists of a nonmetal, carbon, two metalloids, silicon and germanium, and two metals, tin and lead. Each of these elements forms some compounds with formulas which indicate that four other atoms are present per group ⅣA atom, for example, carbon tetrachloride, CCl_4. The group ⅣB metals—titanium, zirconium, and hafnium—also form compounds in which each group ⅣB atom is combined with four other atoms; these compounds are nonelectrolytes when pure.

The elements of group ⅤA include three nonmetals (nitrogen, phosphorus, and arsenic) and two metals (antimony and bismuth). Although compounds with the formulas N_2O_5, PCl_5, and $AsCl_5$ exist, none of them is ionic. These elements do form compounds—nitrides, phosphides, and arsenides—in which ions having charges of minus three occur. The elements of group ⅤB are all metals. These elements form such a variety of different compounds that their characteristics are not easily generalized.

With the exception of polonium, the elements of group ⅥA are typical nonmetals. They are sometimes known as the chalcogens, from the Greek word meaning "ash formers". In their binary compounds with metals they exist as ions having a charge of −2. The elements of group ⅦA are all nonmetals and are known as the halogens from the Greek term meaning "salt formers". They are the most reactive nonmetals and are capable of reacting with practically all the metals and with most nonmetals, including each other.

The elements of groups ⅥB, ⅦB, and ⅧB are all metals. They form such a wide variety of compounds that it is not practical at this point to present any examples as being typical of the behavior of the respective groups.

Reading comprehension

1. What is the characteristic ionic charge of the alkaline earth metals?
2. Why aren't group ⅤB elements' characteristics easily generalized?
3. Which group of elements are known as the halogens?

Lesson two: Inorganic and Organic Compounds

In this lesson you will learn:

- Definition and classification of inorganic compounds
- Definition and classification of organic compounds

1. Introduction of inorganic and organic chemistry

Inorganic chemistry is defined as the study of the chemistry of materials from non-biological origins. Typically, this refers to materials not containing carbon-hydrogen bonds, including metals, salts, and minerals. Inorganic chemistry is used to study and develop catalysts, coatings, fuels, surfactants, materials, superconductors, and drugs. Important chemical reactions in inorganic chemistry include double displacement reactions, acid-base reactions, and redox reactions.

In contrast, chemistry of compounds that contain C—H bonds is called organic chemistry. The organometallic compounds overlap both organic and inorganic chemistry. Organometallic compounds typically include a metal directly bonded to a carbon atom.

surfactant *n.* 表面活性剂

superconductor *n.* 超导体

redox *n.* 氧化还原反应

organometallic *adj.* 有机金属的

M5-7
单词读音及例句

2. Inorganic compounds

Comprising most of the Earth's crust, the class of inorganic compounds is vast and it's difficult to generalize their properties. However, many inorganics are ionic compounds, containing cations and anions joined by ionic bonds. Classes of these salts include oxide, halides, sulfates, and carbonates. Another way to classify inorganic compounds is as main group compounds, coordination compounds, transition metal compounds, cluster compounds, organometallic compounds, solid state compounds, and bioinorganic compounds.

Many inorganic compounds are poor electrical and thermal conductors as solids, have high melting points, and readily assume crystalline structures. Some are soluble in water, while others are not. Usually, the positive and negative electrical charges balance out to form neutral compounds. Inorganic chemicals are common in nature as minerals and electrolytes.

cation *n.* 阳离子

anion *n.* 阴离子

halide *n.&adj.* 卤化物（的）

sulfate *n.* 硫酸盐

carbonate *n.* 碳酸盐

crystalline *adj.* 结晶质的，晶体结构的

minerals *n.* 矿物，矿物质

electrolyte *n.* 电解液，电解质

M5-8
单词读音及例句

M5-9
单词读音及例句

carbon monoxide 一氧化碳
carbide n. 碳化物
carbonate n. 碳酸盐
cyanide n. 氰化物
thiocyanate n. 硫氰酸盐

M5-10
单词读音及例句

It is noteworthy that some simple compounds that contain carbon are often considered inorganic. Examples include carbon monoxide (CO), carbon dioxide (CO_2), carbides, and the salts of carbonates, cyanides and thiocyanates. Many of these are normal parts of mostly organic systems including organisms, hence describing a chemical as inorganic does not necessarily mean that it does not occur within living things.

Notes:

Many of these are normal parts of mostly organic systems including organisms, hence describing a chemical as inorganic does not necessarily mean that it does not occur within living things.

译文：其中许多都是有机系统（包括生物体）的正常组成部分，因此将一个化学物质描述为无机物并不一定意味着它不存在于生物体内。

语法：在副词 hence 引导的分句中，动名词短语 describing... inorganic 作主语，mean 为谓语，连词 that 引导宾语从句。

ammonium nitrate 硝酸铵
fertilizer n. 肥料，化肥
hydrogen n. 氢
nitrogen n. 氮
oxygen n. 氧
valence number 化合价

The first man-made inorganic compound of commercial significance to be synthesized was ammonium nitrate (NH_4NO_3). Ammonium nitrate was made using the Haber process, for use as a soil fertilizer.

3. Organic compounds

Organic compounds are the complex compounds of carbon. Because carbon atoms bond to one another easily, the basis of most organic compounds is comprised of carbon chains that vary in length and shape. Hydrogen, nitrogen, and oxygen atoms are the most common atoms that are generally attached to the carbon atoms. Each carbon atom has 4 as its valence number which increases the complexity of the compounds that are formed. Since carbon atoms are able to form double and triple bonds with other atoms, it also further raises the likelihood for variation in the molecular make-up of organic compounds.

M5-11
单词读音及例句

Notes:

Since carbon atoms are able to form double and triple bonds with other atoms, it also further raises the likelihood for variation in the molecular make-up of organic compounds.

译文：由于碳原子能够与其他原子形成双键和三键，这也进一步提高了有机化合物分子组成产生变化的可能性。

语法：句首的连词 since 引导原因状语从句，意为“因为，由于……”；raise the likelihood for... 意为“提高……的可能性”。

All living things are composed of intricate systems of inorganic and organic compounds. For example, there are many kinds of organic compounds that are found in nature, such as hydrocarbons. Hydrocarbons are the molecules that are formed when carbon and hydrogen combine. They are not soluble in water and easily distribute. There are also aldehydes— the molecular association of a double-bonded oxygen molecule and a carbon atom.

There are many classes of organic compounds. Originally, they were believed to come from living organisms only. However, in the mid-1800s, it became clear that they could also be created from simple inorganic compounds. Yet, many of the organic compounds are associated with basic processes of life, such as carbohydrates, proteins, lipids and nucleic acids.

hydrocarbon *n.* 碳氢化合物，烃类
aldehyde *n.* 醛
carbohydrate *n.* 碳水化合物
protein *n.* 蛋白（质）
lipid *n.* 脂质
nucleic acid *n.* 核酸

M5-12
单词读音及例句

(1) Carbohydrates

Carbohydrates are hydrates of carbon and include sugars. They are quite numerous and fill a number of roles for living organisms. For example, carbohydrates are responsible for storing and transporting energy, maintaining the structure of plants and animals, and helping the functioning of the immune system, blood clotting, and fertilization.

(2) Proteins

Proteins consist of chains of amino acids called peptides. A protein may be made from a single polypeptide chain or may have a more complex structure where polypeptide subunits pack together to form a unit. Proteins consist of hydrogen, oxygen, carbon, and nitrogen atoms. Some proteins contain other atoms, such as sulfur, phosphorus, iron, copper, or magnesium. Proteins serve many functions in cells. They are used to build structure, catalyze biochemical reactions, for immune response, to package and transport materials, and to help replicate genetic material.

immune system 免疫系统
blood clotting 凝血作用
fertilization *n.* 受孕
sulfur *n.* 硫
phosphorus *n.* 磷
iron *n.* 铁
copper *n.* 铜
magnesium *n.* 镁

(3) Lipids

Lipids are made of carbon, hydrogen, and oxygen atoms. Lipids have a higher hydrogen to oxygen ratio than that found in carbohydrates. The three major groups of lipids are triglycerides (fats, oils, waxes), steroids, and phospholipids. Triglycerides consist of three fatty acids joined to a molecule of glycerol. Steroids each have a backbone of four carbon rings joined to each other. Phospholipids resemble triglycerides except there is a phosphate group in place of one of the fatty acid chains. Lipids are used for energy storage, to build structures, and as

triglycerides *n.* 甘油三酯
steroid *n.* 类固醇
phospholipid *n.* 磷脂
resemble *v.* 像，类似……

M5-13
单词读音及例句

M5-14
单词读音及例句

polymer *n.* 聚合物
nucleotide *n.* 核苷酸
phosphate *n.* 磷酸盐

M5-15
单词读音及例句

signal molecules to help cells communicate with each other.

(4) Nucleic acids

A nucleic acid is a type of biological polymer made up of chains of nucleotide monomers. Nucleotides, in turn, are made up of a nitrogenous base, sugar molecule, and phosphate group. Cells use nucleic acids to code the genetic information of an organism.

Reading comprehension

1. What is the first man-made inorganic compound of commercial significance and what is it used for?

2. What is the basis of most organic compounds comprised of?

3. What do cells use nucleic acids for?

Exercise

小提示：
• balance out 译作：相抵，平衡
• 组成：is comprised of
• ……不同：vary in...

1. True or false.

(1) Inorganic chemicals do not occur within living organism.

(2) Organic compounds could be created from simple inorganic molecules.

2. Translate the following sentences into English or Chinese.

(1) Usually, the positive and negative electrical charges balance out to form neutral compounds.____________________

(2) 大多数有机化合物是由长度和形状不同的碳链组成。____________________

volatile *adj.* 挥发性的
define *v.* 给……下定义
solubility *n.* 溶解度
vaporization *n.* 蒸发，气化

Further Reading

What Are Volatile Organic Compounds (VOC)?

The legal definition of volatile organic compounds (VOC) differs from country to country. For example, the U.S. defines volatile organic compounds as organic compounds, basically all hydrocarbons, with low solubility in water, and a propensity for vaporization at relatively low temperatures, including room temperature. Essentially, a volatile organic compound is a chemical or compound that contains such vapor pressure that it does not take much heat to vaporize the particular chemical or

M5-16
单词读音及例句

compound into a gaseous form.

The most common natural volatile organic compound is methane, an indicator of natural gas formation. A common man-made volatile organic compound is formaldehyde, found in furniture components, paint, and many cleaning solutions and disinfectants. Volatile organic compounds and their heretofore unrestricted use have been a major contributor to such environmental issues as smog and "sick building syndrome" .

methane *n.* 甲烷
formaldehyde *n.* 甲醛
disinfectant *n.* 消毒剂，杀菌剂
sick building syndrome 病态建筑综合征

Mainly, the effects of VOC are seen in respiratory ailments, and disorders of the immune system. Increasingly, however, the vaporous effects of volatile organic compounds are recognized as actually being absorbed through the skin, leading to further complications. VOC are now considered more than a nuisance, and they're downright deadly.

respiratory *adj.* 呼吸的
disorder *n.* 混乱，紊乱
nuisance *n.* 损害

The Environmental Protection Agency (EPA), in the mid 1990s, finally established definitions and standards for what constitute dangerous VOC. On 13th September 1999, the first federally mandated regulations for VOC levels, basically EPA volatile organic compounds, went into effect. An EPA volatile organic compound is an organic compound that is determined, by the EPA, to have an especially high photochemical reactivity, or vaporization effect, under normal atmospheric conditions.

mandated *v.* 依法的，强制的
regulation *n.* 规章，条例
photochemical *adj.* 光化学的

What this means is that certain chemicals are especially prone to vaporization under even normal conditions and, therefore, must be kept below certain levels, for instance, industrial coatings, including ordinary house paints, varnishes, *etc*. Most states in the U.S. now have VOC regulations much stricter and more comprehensive than federal statutes.

Improvements have had to be made in the loading and unloading procedures at oil pipeline facilities due to the enormous emissions of volatile organic compounds at these sites (Figure 5-3). Elsewhere, waste sites being cleaned up, buildings re-coated, automobile emissions standards raised, and the chemical processing and manufacturing industry being rigorously regulated, all in an effort to reduce levels of volatile organic compounds.

procedure *n.* 程序，步骤
emission *n.* 排放
automobile *n.* 机动车，汽车

M5-17
单词读音及例句

M5-18
单词读音及例句

M5-19
单词读音及例句

M5-20
单词读音及例句

Figure 5-3 The chemical industry

Reading comprehension

1. What is a volatile organic compound and can you name one VOC?
2. How does VOC affect human health?

Lesson three: Acid and Base

In this lesson you will learn:

- Brønsted theory of acid and base
- pH and pOH
- Acid and base equilibria

propose *v.* 提出（理论）
acidity *n.* 酸性，酸度
alkalinity *n.* 碱性，碱度
donate *v.* 赠送，给予

M5-21
单词读音及例句

1. Brønsted theory of acid and base

In 1923, Brønsted in Denmark and Lowry in England proposed independently a theory of acidity and alkalinity, which is valid in all solvents and is particularly useful in analytical chemistry. The Brønsted-Lowry theory describes acid-base interactions in terms of proton transfer between chemical species. A Brønsted-Lowry acid is any species that can donate a proton, H^+; and a base is any species that can accept a proton. In terms of chemical structure, this means that any Brønsted-Lowry acid must contain a hydrogen that can dissociate as H^+. In order to accept a proton, a Brønsted-Lowry base must have at least one lone pair of electrons to form a new bond with a proton. Acids can be cationic, anionic, or electrically neutral. The same is true for bases. The product formed when an acid gives a proton is called the conjugate base of the parent acid. Similarly, the base produces a conjugate acid when accepting a proton.

Using the Brønsted-Lowry definition, an acid-base reaction is any reaction in which a proton is transferred from an acid to a base. We can use the Brønsted-Lowry definitions to discuss acid-base reactions in any solvent, as well as those that occur in the gas phase. For example, consider the reaction of ammonia gas, $NH_3(g)$, with hydrogen chloride gas, $HCl(g)$, to form solid ammonium chloride, $NH_4Cl(s)$:

$$NH_3(g)+HCl(g) \longrightarrow NH_4Cl(s)$$

This reaction can also be represented using the Lewis structures of the reactants and products, as seen below:

$$\mathrm{H{-}\ddot{N}(H){-}H} \; + \; \mathrm{H{-}Cl} \longrightarrow \mathrm{H{-}\overset{+}{N}(H)(H){-}H} \;\; \mathrm{Cl^-}$$

In this reaction, HCl donates its proton to NH_3. Therefore, HCl is acting as a Brønsted-Lowry acid. Since NH_3 has a lone pair which it uses to accept a proton, NH_3 is a Brønsted-Lowry base.

conjugate acid 共轭酸

conjugate base 共轭碱

ammonia *n.* 氨，氨水

M5-22
单词读音及例句

2. pH and pOH

In aqueous solution, an acid is defined as any species that increases the concentration of $H^+(aq)$, while a base increases the concentration of OH^-. Typical concentrations of these ions in solution can be very small, and they also span a wide range.

For example, a sample of pure water at 25℃ contains 1.0×10^{-7} mol/L of H^+ and OH^-. In comparison, the concentration of H^+ in stomach acid can be up to approximately 1.0×10^{-1} mol/L. That means $[H^+]$ in stomach acid is approximately 6 orders of magnitude larger than that/it in pure water!

To avoid dealing with such cumbersome numbers, scientists convert these concentrations to pH or pOH values.

The pH for an aqueous solution is calculated from $[H^+]$ using eq. (5-1):

$$pH=-lg[H] \tag{5-1}$$

The lowercase p indicates "$-log_{10}$". For example, if we have a solution with $[H^+]=1\times10^{-5}$ mol/L, then we can calculate the pH using eq. (5-2):

$$pH=-lg(1\times10^{-5})=5.0 \tag{5-2}$$

Given the pH of a solution, we can also find $[H^+]$

$$[H^+]=10^{-pH} \tag{5-3}$$

The pOH for an aqueous solution is defined in the same way for $[OH^-]$:

span *v.* 跨越，涵盖

order of magnitude 数量级

convert *v.* 转变，换算

M5-23
单词读音及例句

$$pOH=-\lg[OH^-] \tag{5-4}$$

For example, if we have a solution with $[OH^-]=1\times10^{-12}$ mol/L, pOH can be calculated using eq. (5-5):

$$pOH=-\lg(1\times10^{-12})=12.0 \tag{5-5}$$

Given the pOH of a solution, we can also find $[OH^-]$:

$$10^{-pOH}=[OH^-] \tag{5-6}$$

equilibrium *n.* 平衡

equiliria equilibrium 的复数

M5-24 单词读音及例句

Based on equilibrium concentrations of $[H^+]$ and $[OH^-]$ in water, the following relationship is true for any aqueous solution at 25 ℃:

$$pH+pOH=14 \tag{5-7}$$

This relationship can be used to convert between pH and pOH.

3. Acid-base equilibria

(1) Strong *vs.* weak acids and bases

Strong acids and strong bases refer to species that completely dissociate to form ions in solution. By contrast, weak acids and bases ionize only partially, and the ionization reaction is reversible. Thus, weak acid and base solutions contain multiple charged and uncharged species in dynamic equilibrium.

ionize *vt.&vi.* （使）电离

by contrast 相比之下

dynamic *adj.* 动态的

dissociation constant 电离常数

(2) Weak acids and the acid dissociation constant, K_a

Weak acids are acids that don't completely dissociate in solution. In other words, a weak acid is any acid that is not a strong acid.

The strength of a weak acid depends on how much it dissociates: the more it dissociates, the stronger the acid. In order to quantify the relative strengths of weak acids, we can look at the acid dissociation constant K_a, the equilibrium constant for the acid dissociation reaction.

M5-25 单词读音及例句

For a generic monoprotic weak acid HA, the dissociation reaction in water can be written as follows:

$$HA(aq)+H_2O(l) \rightleftharpoons H_3O^+(aq)+A^-(aq)$$

Based on this reaction, we can write our expression for equilibrium constant K_a as shown in eq. (5-8):

$$K_a=\frac{[H_3O^+][A^-]}{[HA]} \tag{5-8}$$

The equilibrium expression is a ratio of products to reactants. The more HA dissociates into H^+ and the conjugate base A^-, the stronger the acid and the larger the value of K_a. Since pH is related to $[H_3O^+]$, the pH of the solution will be a function of K_a, as well as the concentration of the acid: the pH decreases as the concentration of the acid and/or K_a increase.

Notes:

The more HA dissociates into H^+ and the conjugate base A^-, the stronger the acid and the larger the value of K_a.

译文：HA 解离出越多的 H^+，及其共轭碱 A^-，则其酸性越强，平衡常数 K_a 值越大。

语法：句中 and 前后的 H^+ 和 the conjugate base A^- 为并列结构；另外，句中包含“the+ 比较级，the+ 比较级”的用法，意为“越……，越……”。

(3) Weak bases and the base dissociation constant, K_b

The base dissociation constant, K_b, is also called the base ionization constant. In the ionization reaction for a generic weak base B in water, the base accepts a proton from water to form hydroxide and the conjugate acid, BH^+:

$$B(aq)+H_2O(l) \rightleftharpoons BH^+(aq)+OH^-(aq)$$

The expression for equilibrium constant K_b can be written as follows:

$$K_b=\frac{[BH^+][OH^-]}{[B]} \tag{5-9}$$

The base dissociation constant (or base ionization constant) K_b quantifies the extent of ionization of a weak base. The larger the value of K_b, the stronger the base, and *vice versa*.

quantify *vt.* 量化，确定……的数量

vice versa 反之亦然

Notes:

The base dissociation constant K_b quantifies the extent of ionization of a weak base. The larger the value of K_b, the stronger the base, and *vice versa*.

译文：碱的电离常数 K_b 量化了弱碱的电离化程度。K_b 值越大，碱性越强，反之亦然。

语法：第一句中 of ionization of a weak base 这一介词短语是 extent 的后置定语，其中 of a weak base 又是 ionization 的后置定语；第二句中包含“the+ 比较级，the+ 比较级”的用法，意为“越……，越……”。

M5-26
单词读音及例句

Reading comprehension

1. How does the Brønsted-Lowry theory define an acid and a base？

2. Why do scientists convert $[H^+]$ and $[OH^-]$ concentrations to pH and pOH values?

3. Write the balanced equation for the reaction of hydrogen phosphate acting as a weak base in water.

Exercise

1. True or false.

（1）Weak acids don't completely dissociate in solution.

（2）The smaller the value of K_b, the stronger the base.

（3）pH and pOH can convert between each other.

2. Translate the following sentences into English or Chinese.

（1）The strength of a weak acid depends on how much it dissociates: the more it dissociates, the stronger the acid.____

（2）弱酸和弱碱只部分电离，且电离反应是可逆的。____

小提示：

• The more...the stronger... 比例增减句型，译作"越……越……"。

Further Reading

The pH Scale

Converting $[H^+]$ to pH is a convenient way to gauge the relative acidity or basicity of a solution. The pH scale allows us to easily rank different substances by their pH value.

The pH scale is a negative logarithmic scale. The logarithmic part means that pH changes by 1 unit for every factor of 10 change in concentration of H^+. The negative sign in front of the log tells us that there is an inverse relationship between pH and $[H^+]$. When pH increases, $[H^+]$ decreases, and *vice versa*.

Figure 5-4 shows a pH scale labeled with pH values for some common household substances. These values are for solutions at 25 ℃. Note that it is possible to have a negative pH value.

For a neutral solution, pH=7;

Acidic solutions have pH<7;

Basic solutions have pH>7.

The lower the pH value, the more acidic the solution and the higher

gauge *vt.* 测量，评估

rank *v.* 分等级，排列

logarithmic *adj.* 对数的

M5-27
单词读音及例句

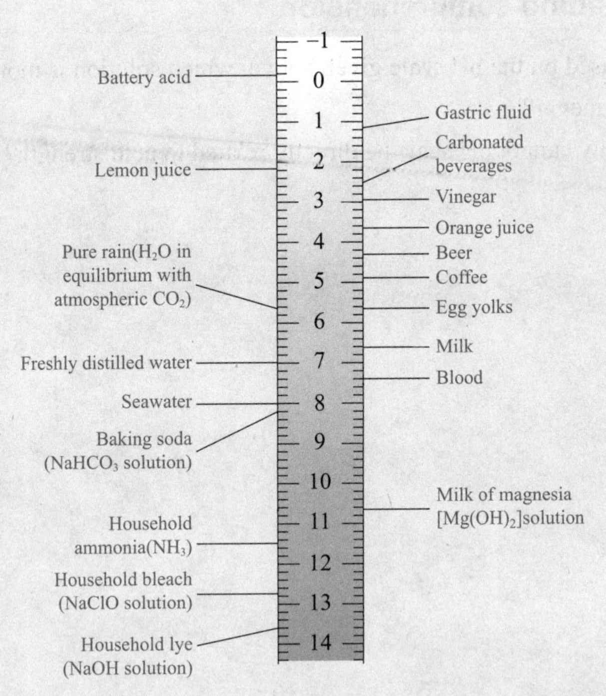

Figure 5-4　pH values for some common household substances

the concentration of H^+. The higher the pH value, the more basic the solution and the lower the concentration of H^+. While we could also describe the acidity or basicity of a solution in terms of pOH, it is more common to use pH.

Relationship between pH and acid strength

Based on the equation for pH, we know that pH is related to $[H^+]$. However, it is important to remember that pH is not always directly related to acid strength. The strength of an acid depends on the amount that the acid dissociates in solution: the stronger the acid, the higher $[H^+]$ at a given acid concentration. For example, a 1.0 mol/L solution of strong acid HCl will have a higher concentration of H^+ than a 1.0 mol/L solution of weak acid HF. Thus, for two solutions of monoprotic acid at the same concentration, pH will be proportional to acid strength.

More generally though, both acid strength and concentration determine $[H^+]$. Therefore, we can't always assume that the pH of a strong acid solution will be lower than the pH of a weak acid solution. The acid concentration also matters.

monoprotic　*adj.*
一元质子的

be proportional to
与……成正比

assume　*v.*
假定，假设

M5-28
单词读音及例句

Reading comprehension

1. Based on the pH scale given above, which solution is more acidic—orange juice, or vinegar?

2. Why cannot pH scale be directly related to acid strength?

第六单元　化学反应与化学平衡

Unit 6　Chemical Reaction and Equilibrium

Lesson one: Introduction to the Laws of Thermodynamics

In this lesson you will learn:

- Laws of thermodynamics
- Principle of thermal equilibrium

1. Introduction

In simplest words, thermodynamics deals with a vast body of understanding of energy transformations and equilibria in all kinds of processes from chemical reactions to windmills. However, a thorough treatment of thermodynamics requires a great deal of mathematics. Sometimes the math seems to obscure the basic ideas and causes confusion. This lesson presents the basic ideas behind the three laws of thermodynamics, uncluttered by equations.

windmill　*n.* 风车

obscure　*vt.* 使……模糊不清

unclutter　*vt.* 使整洁，整理

M6-1
单词读音及例句

2. The 1st law of thermodynamics

The first law is essentially the same as the law of conservation of energy. The energy absorbed or emitted by a system undergoing some kind of process must be accounted for in all the different kinds of energy, *e.g.*, electrical, mechanical, chemical, or heat. Because heat plays an important role in the development of the second law, the different kinds of energy are classified as being either heat, which is energy that flows between bodies at two different temperatures, or work, which encompasses energy that actually causes some kind of motion in large bodies.

conservation of energy 能量守恒

encompass　*vt.* 包括，包含

M6-2
单词读音及例句

Notes:

The different kinds of energy are classified as being either heat, which is energy that flows between bodies at two different temperatures, or work, which encompasses energy that actually causes some kind of motion in large bodies.

译文：不同种类的能量可以分为两类：热量，即在两种不同温度的物体之间传递的能量，或者功，即引起宏观物体进行某种运动的能量。

语法：句中第一个 which 为关系代词，引导非限定性定语从句，修饰 heat, 第二个 which 也为关系代词，引导非限定性定语从句，修饰 work；两个 which 引导的非限定性定语从句中各包含一个由关系代词 that 引导的定语从句，修饰 energy。

intuitively *adv.* 直觉地，直观地

coil *vt.* 将……卷成圈

ignite *vt.&vi.* 点燃，使燃烧

reproducibly *adv.* 重复性地

M6-3
单词读音及例句

The first law then says that:

$$\Delta U = Q + W \tag{6-1}$$

where ΔU is the change in energy of a system, and Q and W are the heat and other kinds of energy added to the system.

While the first law is very useful in tabulating energy changes for various processes, it does not allow one to predict whether or not that process can occur. It simply says that if there is an energy flow, ΔU, it must be accounted for as either heat or work. Yet we are all intuitively aware that the direction of energy flow is governed by some natural law. Water spontaneously flows downhill, a coiled spring unwinds when released, and wood hums unassisted when ignited. These processes reproducibly occur in this way; one never finds, for example, water spontaneously flowing uphill. The law that prescribes the direction of energy flow, and hence the familiar continuity of events, is the second law of thermodynamics.

3. The 2nd law of thermodynamics

cease *v.* 终止，停止

reside *vi.* 驻留，存在

entropy *n.* 熵

criterion *n.* （判断的）标准，判据

spontaneity *n.* 自发性

paradox *n.* 悖论，自相矛盾

M6-4
单词读音及例句

One statement of the second law says that heat will not flow spontaneously between two bodies unless they are at different temperatures. When heat does flow, the hotter body suffers a reduction in temperature, and the cooler body warms up. If heat is allowed to flow between the bodies indefinitely it eventually must cease because the two bodies will come to the same temperature. Although the total energy residing in the two bodies taken together has not changed, something has happened to the bodies, namely, they have lost the ability to exchange heat with each other. Another way of saying this is that their heat has become degraded or useless. This “degradedness” of the two bodies with respect to each other is a kind of quantity just as energy is a quantity, and it has been given the name, entropy. The second law says, that if a spontaneous process is to occur within a system there must be an increase in degradedness or entropy of the entire universe, which consists of the system and its surroundings. This statement goes somewhat beyond our first simple statement in that it does not mention heat. In fact, there need not be an exchange of heat between the system and the surroundings in order that a process is to be spontaneous. All that has to happen is that the total entropy change of system plus surroundings must he positive. Thus, total entropy change for a process is taken as the criterion for spontaneity.

4. The 3rd law of thermodynamics

The third law is the one which allows the evaluation of the absolute entropy of any material, and leads to the interpretation of entropy as the degree of randomness or a state of probability of a system. It says that at absolute zero temperature the absolute entropies of all perfectly crystalline materials are identical and arbitrarily set at 0.

interpretation *n.* 解释，说明

arbitrarily *adv.* 随意地，主观武断的

randomness *n.* 无序性

unequivocal *adj.* 不含糊的，明确的

M6-5
单词读音及例句

Notes:

The third law is the one which allows the evaluation of the absolute entropy of any material, and leads to the interpretation of entropy as the degree of randomness or a state of probability of a system.

译文：第三定律可用于评估任何物质的熵的绝对值，并将熵解释为体系的混乱程度或一种概率状态。

语法：句中 which 为关系代词，引导定语从句，修饰 the one，即 the third law。

The interpretation of entropy as disorder or randomness follows from the recognition that at absolute zero a perfect crystal is perfectly ordered, *i.e.*, if one were to draw a three-dimensional picture of such a crystal he would have to produce only one unequivocal picture of it. At temperatures higher than absolute zero, the atoms in the crystal vibrate in excited states, imperfections occur, and eventually the lattice breaks up. The crystal becomes more and more disordered and the number of pictures of how the crystal might look becomes larger and larger. The bridge between classical thermodynamics and statistics is made by postulating that the entropy of any system is determined by the number of ways of arranging the particles in that system without changing the state. In other words, entropy is directly related to probability. Thus, natural processes always occur such that the universe evolves from a state of lower to a state of higher probability.

Reading comprehension

1. What does the 1st law of thermodynamics state?
2. What is the criterion for spontaneity according to the 2nd law?
3. What happens to the absolute entropies of all perfectly crystalline materials at absolute zero?

Exercise

1. March the following Chinese and English.

conservation of energy	能量守恒
spontaneous process	动能
energy conversion	激发态
kinetic energy	能量转化
excited state	自发过程

2. Translate the following sentences into English or Chinese.

（1）One statement of the second law says that heat will not flow spontaneously between two bodies unless they are at different temperatures.____________

____

（2）当热量流动时，较热的物体温度降低，较冷的物体升温。______

____

小提示：

• not...unless... 否定条件的状语从句，译作“不……除非……”。

• hotter: hot 的比较级，较热的

cooler: cool 的比较级，较冷的

Further Reading

The History of Thermometer

Temperature is a broad concept, which has the closest relationship with each of us. The comfortable body temperature is about 37 ℃. If the body temperature rises or decreases, it will cause physical discomfort, so that you can feel the “severe temperature” personally.

In fact, in the process of human production and life, the concept of temperature was everywhere, such as smelting and forging according to temperature, spring ploughing and winter storage, and even affecting the victory or defeat of wars. So how did people measure the temperature in the ancient time?

Let’s take a look at the thermometer in the history of modern science. According to historical records, Galileo, an Italian scientist, invented the world’s first thermometer in 1653. Using the principle of thermal expansion and cold contraction, he injected red liquid into a slender graduated glass tube, inserted one end into the sink, and the other end of the glass tube was a glass ball as shown in Figure 6-1 (a). Therefore, if the indoor temperature rises, the air in the glass ball will expand and push the red liquid in the glass tube downward, so as to confirm the temperature according to the change of scale.

In the Qing Dynasty, the missionary Nan Huairen（南怀仁）invented a more accurate thermometer based on Galileo’s model. Although the principle was similar, Nan Huairen replaced the slender glass tube with

discomfort *n. & vt* 不舒服，使不舒服

severe *adj.* 十分严重的

plough *vt. & n.* 耕作，耕地

defeat *v. & n.* 击败，战败

ancient *adj.* 古代的，古老的

M6-6
单词读音及例句

a U-tube [Figure 6-1 (b)]. At the same time, the scale was further refined to make the measurement results more accurate. Since then, the concept of "thermometer" was brought to China.

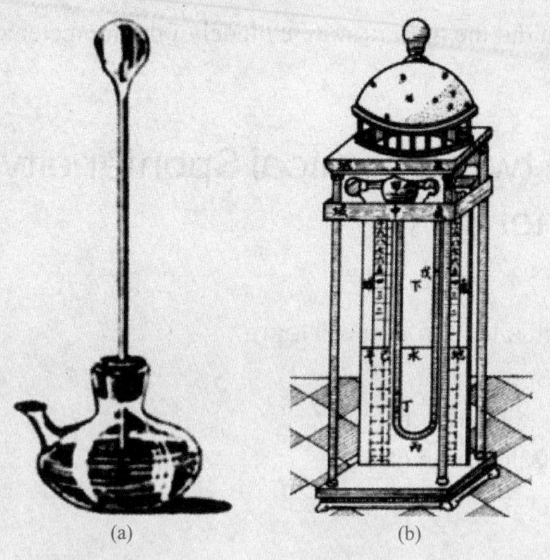

(a)　　(b)

Figure 6-1　Galileo's (a) and Nan Huairen's (b) model of thermometer

In fact, our ancient ancestors had the consciousness of measuring temperature as early as the pre-Qin period. It was first recorded in *Lü's Commentaries of History*（《吕氏春秋》）, saying that "when you see ice in the water bottle, you know that the world is cold, and the fish and turtles hide", which means that as long as you see that the water in the bottle freezes, the world will enter a cold season, and the animals are ready to hide and hibernate.

After in-depth research, the archaeologists identified that the "bottle" mentioned in the literature, namely the "ice bottle", was an instrument for measuring temperature at that time, and it was also the earliest thermometer in the world. By putting clear water into the bottle and observing the change of water, one can tell whether it is cold or warm. It has shown that our ancestors had the concept of "temperature" as early as two or three thousand years ago. However, the historical records on the "ice bottle" did not specify its size and shape, causing difficulties to identify it during archaeological excavation.

Although simple and primitive, the "ice bottle" was fully functional. Through careful observation of the nature, our ancestors integrated their experience in life and made useful inventions, not only having improved the agricultural work but also promoted the development of human history.

missionary　*n.* 传教士

slender　*adj.* 苗条的

refine　*vt.* 提炼，改善

M6-7 单词读音及例句

ancestor　*n.* 祖先

consciousness　*n.* 意识

hibernate　*vi.* 冬眠

M6-8 单词读音及例句

archaeologist　*n.* 考古学家

excavation　*n.* 发掘

primitive　*adj.* 原始的

agricultural　*adj.* 农业的

M6-9 单词读音及例句

Reading comprehension

1. What is the principle that Galileo used to invent the first thermometer in modern science history?

2. When did the most primitive model of thermometer appear in China?

Lesson two: Chemical Spontaneity and Gibbs Free Energy

In this lesson you will learn:

- Spontaneous processes
- Gibbs free energy
- Calculation of ΔG

1. Spontaneous processes

graphite　*n.* 石墨

detectable　*adj.* 可发觉的

timescale　*n.* 时间量程

M6-10
单词读音及例句

In chemistry, a spontaneous process is one that occurs without the addition of external energy. A spontaneous process may take place quickly or slowly, because spontaneity is not related to kinetics or reaction rate. A classic example is the process of carbon in the form of a diamond turning into graphite, which can be written as the following reaction:

$$C(s,diamond) \rightarrow C(s,graphite)$$

This reaction takes so long that it is not detectable on the timescale of (ordinary) humans, hence the saying, "diamonds are forever." If we could wait long enough, we should be able to see carbon in the diamond form turns into the more stable but less shiny, graphite form.

diamond

graphite

Notes:

If we could wait long enough, we should be able to see carbon in the diamond form turn into the more stable but less shiny, graphite form.

译文：如果我们能等待足够长的时间，我们应该能够看到金刚石形式的碳变成更稳定但不那么闪亮的石墨形式。

语法：本句使用虚拟语气，是连词 if 引导的非真实条件句，表示与现在事实相反（事实上人类无法等待足够长的时间，无法看到金刚石形式的碳变成石墨形式）。

Another thing to remember is that spontaneous processes can be exothermic or endothermic. That is another way of saying that spontaneity is not necessarily related to the enthalpy change of a process, ΔH.

How do we know if a process will occur spontaneously? The short but slightly complicated answer is that we can use the second law of thermodynamics. According to the second law of thermodynamics, any spontaneous process must increase the entropy in the universe. This can be expressed mathematically as follows:

$$\Delta S_{universe}=\Delta S_{system}+\Delta S_{surroundings}>0 \quad (6\text{-}2)$$

Hence one has to do is measure the entropy change of the whole universe. Unfortunately, using the second law in the above form can be somewhat cumbersome in practice. After all, most of the time chemists are primarily interested in changes within our system, which might be a chemical reaction in a beaker. Luckily, chemists can get around having to determine the entropy change of the universe by defining and using a new thermodynamic quantity called Gibbs free energy.

cumbersome *adj.* 繁琐的

beaker *n.* 烧杯

Gibbs free energy 吉布斯自由能

get around 绕开

M6-11
单词读音及例句

2. Gibbs free energy

When a process occurs at constant temperature T and pressure P, we can rearrange the second law of thermodynamics and define a new quantity known as Gibbs free energy:

$$\text{Gibbs free energy}=G=H-TS \quad (6\text{-}3)$$

where H is enthalpy, T is temperature (in kelvin, K), and S is the entropy. Gibbs free energy is represented using the symbol G and typically has units of kJ/mol.

enthalpy *n.* 焓

entropy *n.* 熵

reactant *n.* 反应物

M6-12
单词读音及例句

When using Gibbs free energy to determine the spontaneity of a process, we are only concerned with changes in G, rather than its absolute value. The change in Gibbs free energy for a process is thus written as ΔG, which is the difference between G_{final}, the Gibbs free energy of the products, and $G_{initial}$, the Gibbs free energy of the reactants.

$$\Delta G=G_{final}-G_{initial} \quad (6\text{-}4)$$

For a process at constant T and constant P, we can rewrite the equation for Gibbs free energy in terms of changes in the enthalpy (ΔH_{system}) and entropy (ΔS_{system}) for our system:

$$\Delta G_{system}=\Delta H_{system}-T\Delta S_{system} \quad (6\text{-}5)$$

subscript *n.* 下标，脚注

surrounding *n.* 周围

M6-13
单词读音及例句

You might also see this reaction written without the subscripts specifying that the thermodynamic values are for the system (not the surroundings or the universe), but it is still understood that the values for ΔH and ΔS are for the system of interest. This equation is exciting because it allows us to determine the change in Gibbs free energy using the enthalpy change, ΔH, and the entropy change, ΔS, of the system. We can use the sign of ΔG to figure out whether a reaction is spontaneous in the forward direction, backward direction, or if the reaction is at

exergonic *adj.* 放能的

endergonic *adj.* 吸能的

constant *n. & adj.* 常量，恒定的

M6-14
单词读音及例句

equilibrium.

- When $\Delta G<0$, the process is exergonic and will proceed spontaneously in the forward direction to form more products.
- When $\Delta G>0$, the process is endergonic and not spontaneous in the forward direction. Instead, it will proceed spontaneously in the reverse direction to make more starting materials.
- When $\Delta G=0$, the system is in equilibrium and the concentrations of the products and reactants will remain constant.

3. Calculating change in Gibbs free energy

assume *v.* 假设

phase *n.* （化学）相

estimate *vt.* 估计，估算

M6-15
单词读音及例句

Although ΔG is temperature dependent, it's generally alright to assume that the ΔH and ΔS are independent of temperature as long as the reaction does not involve a phase change. That means that if we know ΔH and ΔS, we can use those values to calculate ΔG at any temperature. The values of ΔH and ΔS can be calculated using many methods including estimating $\Delta H_{reaction}$ using bond enthalpies; calculating ΔH using standard heats of formation, $\Delta_f H^o$ and calculating ΔH and ΔS using tables of standard values. When the process occurs under standard conditions [all gases at 1bar (1 bar=10^5 Pa) pressure, all concentrations are 1 mol/L, and T=25 ℃], one can also calculate ΔG using the standard free energy of formation, $\Delta_f G^o$.

Notes:

Although ΔG is temperature dependent, it's generally alright to assume that the ΔH and ΔS are independent of temperature as long as the reaction does not involve a phase change.

译文：尽管 ΔG 与温度有关，但只要反应不涉及相变，一般可以假设 ΔH 和 ΔS 与温度无关。

语法：句首 although 为连词，引导让步状语从句；that 为连词，引导宾语从句；as long as 引导条件状语从句，意为“只要”。

We can make the following conclusions about when processes will have a negative ΔG_{system}:

- When the process is exothermic ($\Delta H_{system}<0$), and the entropy of the system increases ($\Delta S_{system}>0$), the sign of ΔG_{system} is negative at all temperatures. Thus, the process is always spontaneous.
- When the process is endothermic, ($\Delta H_{system}>0$) and the entropy of the system decreases, $\Delta S_{system}<0$, the sign of ΔG is positive at all temperatures. Thus, the process is never spontaneous.

For other combinations of ΔH_{system} and ΔS_{system}, the spontaneity of a process depends on the temperature.

Calculating ΔG for melting ice

Let's consider an example that illustrates the effect of temperature on the spontaneity of a process. The enthalpy of fusion and entropy of fusion for water have the following values:

$$\Delta_{fus}H = 6.01 kJ/mol$$

$$\Delta_{fus}S = 22.0 J/(mol \cdot K)$$

What is ΔG for the melting of ice at 20 ℃ ?

The process we are considering is water changing phase from solid to liquid:

$$H_2O(s) \longrightarrow H_2O(l)$$

If we plug the values for ΔH, T and ΔS into the following equation:

$$\Delta G = \Delta H - T\Delta S = 6.01-(20+273)\times 0.022 = -0.44(kJ/mol)$$

Since ΔG is negative, we would predict that ice spontaneously melts at 20 ℃ .

fusion　*n.* 熔化

M6-16
单词读音及例句

Reading comprehension

1. What is a spontaneous process?

2. How to determine whether a reaction is spontaneous in terms of Gibbs free energy?

3. Can you think of any processes in your day-to-day life that are spontaneous at certain temperatures but not at others?

Exercise

1. True or false.

(1) ΔG is temperature independent.

(2) All spontaneous process must increase the entropy in the universe.

(3) The process is never spontaneous, when it is endothermic and the entropy of the system decreases.

2. Translate the following sentences into English or Chinese.

（1）When using Gibbs free energy to determine the spontaneity of a process, we are only concerned with changes in G, rather than its absolute value.________

____

____

（2）ΔG 的符号可用于判断一个化学反应是会向正向自发进行，逆向自发进行，还是处于平衡状态。________________________

____

小提示：

• be concerned with 分词做形容词，译作“考虑……”。

• forward direction: 正向

backward direction: 逆向

Further Reading

convert *v.* （使）转变

empirical *adj.* 经验主义的

deductive *adj.* 推导的，演绎的

pursue *vt&vi.* 追求，学习

thesis *n.* 论文

be distinguished by 以……为特征

devote *v.* 奉献，把……用于

M6-17 单词读音及例句

Profiles in Chemistry—J. Willard Gibbs (1839—1903)

J. Willard Gibbs, in full Josiah Willard Gibbs, (born February 11, 1839, New Haven, Connecticut, U.S.—died April 28, 1903, New Haven), was a theoretical physicist and chemist who was one of the greatest scientists in the United States in the 19th century. His application of thermodynamic theory converted a large part of physical chemistry from an empirical into a deductive science.

Gibbs was the fourth child and only son of Josiah Willard Gibbs, Sr., professor of sacred literature at Yale University. There were college presidents among his ancestors and scientific ability in his mother's family. Facially and mentally, Gibbs resembled his mother. He was a friendly youth but was also withdrawn and intellectually absorbed. This circumstance and his delicate health kept him from participating much in student and social life. He was educated at the local Hopkins Grammar School and in 1854 entered Yale, where he won a succession of prizes. After graduating, Gibbs pursued research in engineering. His thesis on the design of gearing was distinguished by the logical rigour with which he employed geometrical methods of analysis. In 1863 Gibbs received the first doctorate of engineering to be conferred in the United States. He was appointed a tutor at Yale in the same year. He devoted some attention to engineering invention.

intellectual *adj.* 智力的

assimilate *v.* 同化，融入

by means of 依靠，借助于

M6-18 单词读音及例句

Gibbs lost his parents rather early, and he and his two older sisters inherited the family home and a modest fortune. In 1866 they went to Europe, remaining there nearly three years while Gibbs attended the lectures of European masters of mathematics and physics, whose intellectual technique he assimilated. He returned more a European than an American scientist in spirit—one of the reasons why general recognition in his native country came so slowly. He applied his increasing command of theory to the improvement of James Watt's steam-engine governor. In analyzing its equilibrium, he began to develop the method by which the equilibriums of chemical processes could be calculated. He was appointed professor of mathematical physics at Yale in 1871, before he had published his fundamental work. His first major paper was "Graphical Methods in the Thermodynamics of Fluids," which appeared in 1873. It was followed in the same year by "A Method of Geometrical Representation of the Thermodynamic Properties of Substances by Means of Surfaces" and in 1876 by his most famous paper, "On the Equilibrium of Heterogeneous Substances". The importance of his work was immediately recognized by the Scottish physicist James Clerk Maxwell in England, who constructed a model of Gibbs's thermodynamic surface with his own hands and sent it to him.

Gibbs was highly esteemed by his friends, but U.S. science was too preoccupied with practical questions to make much use of his profound theoretical work during his lifetime. He lived out his quiet life at Yale, deeply admired by a few able students but making no immediate impress on U.S. science commensurate with his genius. He never even became a member of the American Physical Society. He seems to have been unaffected by this. He was aware of the significance of what he had done and was content to let posterity appraise him.

profound *adj.* 深厚的

admire *v.* 钦佩，赞赏

appraise *v.* 评价，评估

M6-19
单词读音及例句

The contemporary historian Henry Adams called Gibbs "the greatest of Americans, judged by his rank in science". His application of thermodynamics to physical processes led him to develop the science of statistical mechanics; his treatment of it was so general that it was later found to apply as well to quantum mechanics as to the classical physics from which it had been derived.

Reading comprehension

1. When did Gibbs publish his most famous paper and what is the title of this paper?

2. How did historian Henry Adams praise Gibbs and why?

Lesson three: Chemical Equilibrium

In this lesson you will learn:

- Definition of equilibrium
- Equilibrium constant
- Le Chatelier's principle

equilibrium *n.* 平衡

net change 净变化

dynamic *adj.* 动态的

M6-20
单词读音及例句

1. Definition of equilibrium

Chemical equilibrium is the condition which occurs when the concentrations of reactants and products participating in a chemical reaction exhibit no net change over time. This does not mean the chemical reaction has stopped occurring, but that the consumption and formation of substances have reached a balanced condition. The quantities of reactants and products have achieved a constant ratio, but they are not necessarily equal. In other words, the forward rate of reaction equals the backward rate of reaction. Chemical equilibrium is also known as **dynamic equilibrium**.

reversible *adj.* 可逆的

irreversible *adj.* 不可逆的

evacuate *v.* 排空，疏散

M6-21
单词读音及例句

Most reactions are theoretically reversible in a closed system, though some can be considered to be irreversible if they heavily favor the formation of reactants or products. One example of a reversible reaction is the formation of nitrogen dioxide, NO_2 from dinitrogen tetroxide, N_2O_4:

$$N_2O_4(g) \rightleftharpoons 2NO_2(g)$$

Imagine we added some colorless $N_2O_4(g)$ to an evacuated glass container at room temperature. If we kept our eye on the vial over time, we would observe the gas in the ampoule changing to a yellowish orange color and gradually getting darker until the color stayed constant. We can graph the concentration of NO_2 and N_2O_4 over time for this process, as you can see in Figure 6-2.

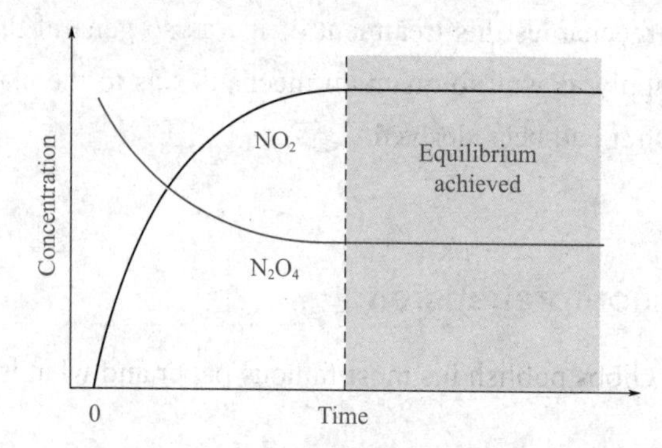

Figure 6-2 Concentration over time for conversion between N_2O_4 and NO_2

Initially, the vial contains only N_2O_4, and the concentration of NO_2 is 0 mol/L. As N_2O_4 gets converted to NO_2, the concentration of NO_2 increases up to a certain point, indicated by a dotted line in the graph to the left, and then stays constant. Similarly, the concentration of N_2O_4 decreases from the initial concentration until it reaches the equilibrium concentration. When the concentrations of NO_2 and N_2O_4 remain constant, the reaction has reached equilibrium.

All reactions tend towards a state of chemical equilibrium, the point at which both the forward process and the reverse process are taking place at the same rate. Since the forward and reverse rates are equal, the concentrations of the reactants and products are constant at equilibrium. It is important to remember that even though the concentrations are constant at equilibrium, the reaction is still happening! That is why this state is also sometimes referred to as dynamic equilibrium.

Notes:

All reactions tend towards a state of chemical equilibrium, the point at which both the forward process and the reverse process are taking place at the same rate.

译文：所有的反应都趋向于化学平衡状态，即正反应和逆反应以同样的速率进行。

语法：句中 at which 引导定语从句，关系代词 which 指代先行词 the point。

2. Equilibrium constant

equilibrium constant 平衡常数

subscript *n.* 下标

molar concentration 摩尔浓度

M6-22 单词读音及例句

Based on the concentrations of all the different reaction species at equilibrium, we can define a quantity called the equilibrium constant K_c, which is also sometimes written as K_{eq}. The c in the subscript stands for concentration since the equilibrium constant describes the molar concentrations at equilibrium for a specific temperature. The equilibrium constant can help us understand whether the reaction tends to have a higher concentration of products or reactants at equilibrium. We can also use K_c to determine if the reaction is already at equilibrium.

Consider the balanced reversible reaction below:

$$a\mathrm{A}+b\mathrm{B} \rightleftharpoons c\mathrm{C}+d\mathrm{D}$$

If we know the molar concentrations for each reaction species, we can find the value for K_c using the relationship shown in eq (6-6):

$$K_c=\frac{[\mathrm{C}]^c[\mathrm{D}]^d}{[\mathrm{A}]^a[\mathrm{B}]^b} \tag{6-6}$$

where [C] and [D] are equilibrium product concentrations; [A] and [B] are equilibrium reactant concentrations; and a, b, c and d are the stoichiometric coefficients from the balanced reaction.

stoichiometric coefficient 化学计量系数

balanced *adj.* 配平的

M6-23 单词读音及例句

There are some important things to remember when calculating K_c:

- K_c is a constant for a specific reaction at a specific temperature. If you change the temperature of a reaction, then K_c also changes.
- Pure solids and pure liquids, including solvents, are not included in the equilibrium expression.
- The reaction must be balanced with the coefficients written as the lowest possible integer values in order to get the correct value K_c.
- If any of the reactants or products are gases, we can also write the equilibrium constant in terms of the partial pressure of the gases.

The magnitude of K_c can give us some information about the reactant and

guideline *n.* 方针，原则

estimate *vt.* 估算，判断

favor *v.* 有利于

M6-24
单词读音及例句

product concentrations at equilibrium:

- If K_c is very large, ~1000 or more, there will be mostly product species present at equilibrium.
- If K_c is very small, ~0.001 or less, there will be mostly reactant species present at equilibrium.
- If K_c is between 0.001 and 1000, there will be a significant concentration of both reactant and product species present at equilibrium.

By using these guidelines, one can quickly estimate whether a reaction will strongly favor the forward direction to make products or strongly favor the backward direction to make reactants.

3. Le Chatelier's principle

Le Chatelier's principle 勒夏特列原理

readjust *vt.&vi.* 再调整

counter *v.* 抵消，抵制

M6-25
单词读音及例句

Le Chatelier's principle can be used to predict the shift in equilibrium resulting from applying a stress to the system. The principle is named for Henry Louis Le Chatelier and is also known as the equilibrium law. The law may be stated:

When a system at equilibrium is subjected to a change in temperature, volume, concentration, or pressure, the system readjusts to partially counter the effect of the change, resulting in a new equilibrium.

Notes:

When a system at equilibrium is subjected to a change in temperature, volume, concentration, or pressure, the system readjusts to partially counter the effect of the change, resulting in a new equilibrium.

译文：当处于平衡状态的系统受到温度、体积、浓度或压力变化的影响时，系统会重新调整以部分抵消变化的影响，从而产生新的平衡。

语法：句首 when 为连词，引导时间状语从句；句尾 resulting in a new equilibrium 为现在分词作状语。

externally *adv.* 从外部

exothermic *adj.* 放热的

M6-26
单词读音及例句

(1) Concentration

An increase in the amount of reactants (their concentration) will shift the equilibrium to produce more products (product-favored). Increasing the number of products will shift the reaction to make more reactants (reactant-favored). Decreasing reactants favors reactants. Decreasing products favors products.

(2) Temperature

Temperature may be added to a system either externally or as a result of the chemical reaction. Increasing or decreasing temperature can be considered the same as increasing or decreasing the concentration of reactants or products. For an exothermic reaction (generate heat), if the temperature is increased, the heat of the system increases, causing the equilibrium to shift to the left (reactants). If the temperature is decreased, the equilibrium shifts to the right (products). In other words, the system compensates for the reduction in temperature by favoring the reaction that generates heat.

(3) Pressure/Volume

Pressure and volume can change if one or more of the participants in a chemical reaction is a gas. Changing the partial pressure or volume of a gas acts the same as changing its concentration. If the volume of gas increases, pressure decreases (and *vice versa*). If the pressure or volume increases, the reaction shifts toward the side with lower pressure. Note, however, that adding an inert gas (*e.g.*, argon or neon) increases the overall pressure of the system, yet does not change the partial pressure of the reactants or products, so no equilibrium shift occurs.

In addition to chemistry, the principle also applies, in slightly different forms, to the fields of pharmacology and economics.

inert　*adj.* 惰性的
argon　*n.* 氩气
neon　*n.* 氖气
pharmacology　*n.* 药理学
economics　*n.* 经济学

M6-27
单词读音及例句

Reading comprehension

1. How to describe a chemical reaction which is at equilibrium?
2. What information does the magnitude of equilibrium constant K_c tell us?
3. What does Le Chatelier's principle state?

Exercise

1. True or false.

（1）Most reactions are theoretically reversible in a closed system.

（2）For the reaction with equilibrium constant K_c of 1, there will be mostly product species present at equilibrium.

（3）Changing the partial pressure of a gas has the same effect on equilibrium as changing its concentration.

2. Translate the following sentences into English or Chinese.

（1）The quantities of reactants and products have achieved a constant ratio, but they are not necessarily equal.____
____
____

（2）勒夏特列原理可用于预测可逆反应的平衡移动。________________
____

小提示：
• constant ratio 译作：恒定的比率，定比
• “平衡移动”英文表达：shift in equilibrium

Further Reading

remarkable *adj.* 显著的

polymer *n.* 聚合物

hydrocarbon *n.* 烃类

M6-28
单词读音及例句

Rubber Bands, Free Energy, and Le Châtelier's Principle

Background

All of us use rubber bands. This activity reveals some remarkable properties of these familiar objects. Rubber molecules are polymers with long hydrocarbon chains. As illustrated in Figure 6-3, the elasticity of rubber is a consequence of the different degrees of order of these chains in the stretched and unstretched states.

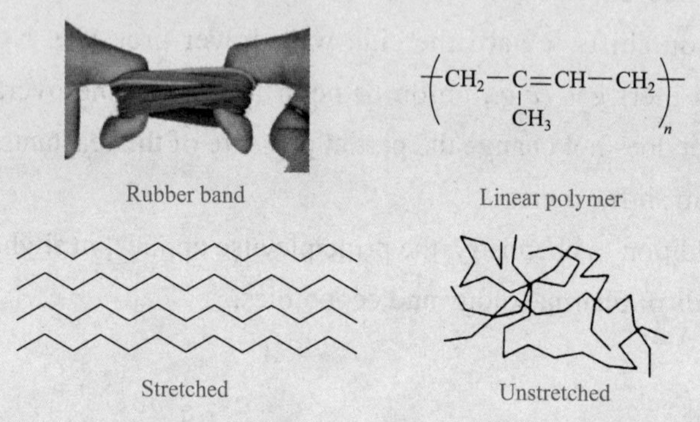

Figure 6-3 The molecular structure and configeration of rubber band

configuration *n.* 构造，结构

revert *vi.* 恢复，重回

align *v.* 对齐，对准，排列

M6-29
单词读音及例句

The stretched configuration is more ordered—of lower entropy. When released, it will revert to the more disordered, higher entropy, unstretched state. Therefore, ΔS, the entropy change, is negative for stretching. In the stretched state, the molecules are better aligned and there are stronger intermolecular attractions. Since formation of such attractions is exothermic, there is heat transfer to the surroundings when rubber is stretched. That is, stretching has a negative enthalpy change, ΔH.

Procedure of the experiment (Figure 6-4)

Try This

You will need: large rubber band about 8 cm long by 0.5 cm wide; strip 10 cm long by 0.5 cm wide cut from a similar rubber band; scissors; lamp with high-intensity incandescent bulb; digital balance accurate to at least 0.1 g; ring stand; clamp fastener; weight with a mass of about 150 g (an ordinary combination lock will do if such a weight isn't available); clock.

1. Touch an unstretched rubber band against your forehead for a few seconds and then remove it. Now hold it close to your forehead, stretch it until it is about three times its original length, and *quickly* touch it to your forehead. Note any temperature change. Repeat the steps several times, if necessary, until you are certain of the results.
2. Stretch a rubber band so that it is about three times its original length and hold it in the air for about 20 seconds. Now let it contract carefully and *quickly* touch it to your forehead. Note any temperature change.
3. Using Le Châtelier's principle, predict what effect the stress of heating will have on a stretched rubber strip.
4. Cut a strip 10 cm long by 0.5 cm from a rubber band. Tie one end to a clamp fastener and the other to a weight that has a mass of about 150 g. Attach the clamp fastener to a ring stand. Adjust the height of the clamp so that the weight rests on an electronic balance, with a registered mass of approximately 20 g. When the balance reading stabilizes, record it.

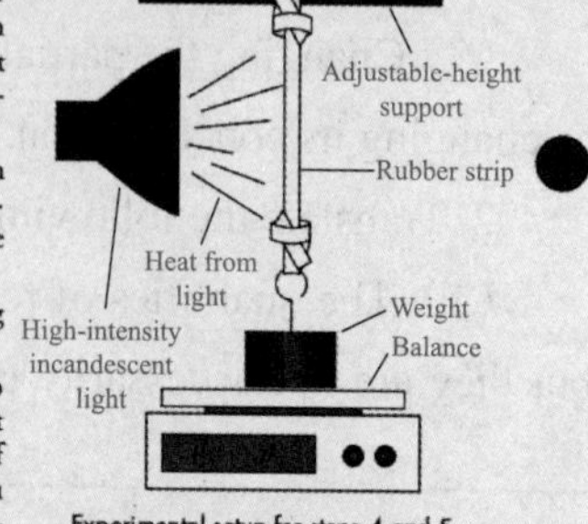

Experimental setup for steps 4 and 5

5. Bring a high-intensity incandescent light bulb as close as possible to the rubber strip without touching and turn it on. Record any mass changes for about 45 seconds. Do your observations support your prediction in step 3? How?

Figure 6-4 The rubber band experiment setup and procedure

Results and discussion

A rubber band that is stretched and quickly touched to the forehead feels noticeably warm. When the stretched band is held in the air for a few seconds, allowed to contract, and then quickly touched to the forehead, there is a noticeable cooling. A rubber strip stretched by means of a mass of about 150 g is gently seated on an electronic balance so that about 20 g of its mass is registered. When a high intensity lamp warms the strip, an apparent mass decrease of several tenths of a gram occurs in the course of about 45 seconds. This indicates that the stretched rubber contracts upon heating.

Le Châtelier's principle concerning the effect of stresses on equilibria is a useful tool in predicting changes. Based on the above discussion, it is established that stretching a rubber strip is exothermic and unstretching is endothermic. According to Le Châtelier's principle, when a stress (heat in this case) is applied to a system in equilibrium, the equilibrium will shift to the side which is endothermic so as to relieve the stress (heat in this case). Hence heating a stretched rubber band causes it to shrink.

forehead *n.* 前额

noticeably *adv.* 显著地

contract *v.* 收缩

M6-30
单词读音及例句

register *v.* 显示，注册

apparent *adj.* 明显的

equilibria equilibrium 的复数

M6-31
单词读音及例句

Reading comprehension

1. How does the entropy change when a rubble band is stretched?

2. Explain why heating a stretched rubber band causes it to shrink using Le Châtelier's principle.

Lesson four: Reaction Rates

In this lesson you will learn:

- Reaction rates and expressions
- Reaction laws
- Reaction orders

1. Introduction

Chemical kinetics is a branch of physical chemistry that is concerned with understanding the rates of chemical reactions. It is to be contrasted with thermodynamics, which deals with the direction in which a process occurs but in

kinetics *n.* 动力学

cosmology *n.* 宇宙学

geology *n.* 地质学

far-reaching 影响深远的

M6-32
单词读音及例句

itself tells nothing about its rate. Thermodynamics is time's arrow, while chemical kinetics is time's clock. Chemical kinetics relates to many aspects of cosmology, geology, biology and engineering, and thus has far-reaching implications. The principles of chemical kinetics apply to purely physical processes as well as to chemical reactions.

One reason why kinetics is important is that it provides evidence for the mechanisms of chemical processes. Besides being of intrinsic scientific interest, knowledge of reaction mechanisms is of practical use in deciding what is the most effective way of causing a reaction to occur. Many commercial processes can take place by alternative reaction paths, and knowledge of the mechanisms makes it possible to choose reaction conditions that favour one path over others.

Notes:

It is to be contrasted with thermodynamics, which deals with the direction in which a process occurs but in itself tells nothing about its rate.

译文：其与热力学的区别在于，热力学只研究一个过程发展的方向，但无法体现发展的速率。

语法：句首 it 指 chemical kinetics；句中第一个 which 为关系代词，指代 thermodynamics, 引导非限制性定语从句；in which 引导定语从句，修饰 direction。

2. The rates of reactions

instantaneous *adj.* 瞬时的

consumption *n.* 消耗，消费

tangent *n.* （数）正切

M6-33
单词读音及例句

The rates of reactions are usually defined as the change of reactant and product concentration per unit time. Providing the volume of the system is constant, the instantaneous rate of consumption of a reactant or formation of a product is the slope of the tangent to the graph of concentration against time as shown in Figure 6-5.

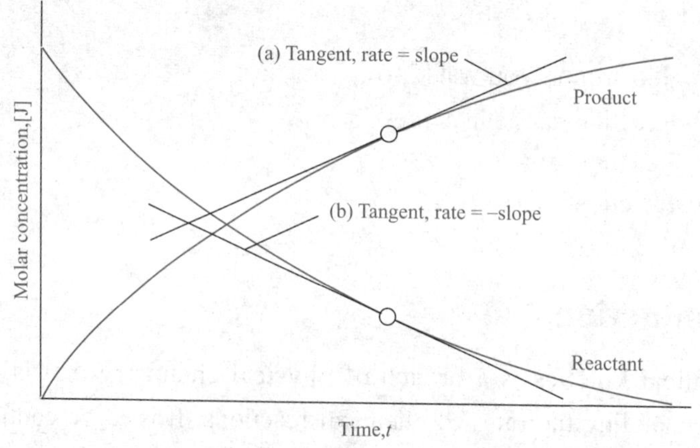

Figure 6-5 Concentration of (a) products, (b) reactants against time

However, as the stoichiometric number of different species may vary, there are several rates connected with the reaction. The undesirability of having different rates to describe the same reaction is avoided by introducing the **extent of reaction**, ξ, which is defined so that for each species J in the reaction, the change in amount of J, dn_J, is

$$d\xi = dn_J/v_J \tag{6-7}$$

where v_J is the stoichiometric number of the species. As the reaction proceeds, the extent of reaction increases. In physical chemistry, the unique rate of reaction, v, is expressed as variation of extent of reaction with time in unit volume:

$$v = (1/V) \times (d\xi /dt) \tag{6-8}$$

where V is the volume of the system. For any species J, $d\xi = dn_J/v_J$, so

$$v = (1/V) \times dn_J/(v_J dt) \tag{6-9}$$

For a homogeneous reaction in a constant-volume system the volume V can be taken inside the differential and n_J/V is written as the molar concentration to give:

$$v = dc_J/(v_J dt) \tag{6-10}$$

Where c_J is the molar concentration of J, mol/L; v_J is the stoichiometric number of J. dc_J/dt represents the variation of the concentration of species against time. To reactants, dc_J/dt and v_J are both negative values, whereas to products, both values are positive. Hence the rate of reactions is always positive regardless of which species is used to represent.

Consider a reaction of the form $a\text{A} + d\text{D} \longrightarrow e\text{E} + f\text{F}$ reaction rate v is related to the stoichiometry and can be expressed as follows:

$$v= (-1/a)(dc_A/dt) = (-1/d)(dc_D/dt) = (1/e)(dc_E/dt) = (1/f)(dc_F/dt) \tag{6-11}$$

It is usually more convenient to describe the reaction progress using the consumption rate of a reactant or the formation rate of a product. Providing the volume is constant, the consumption rate of a reactant can be written as:

$$v_A=-dc_A/dt \tag{6-12}$$

$$v_B=-dc_B/dt \tag{6-13}$$

The formation rate of a product can be written as:

$$v_E=dc_E/dt \tag{6-14}$$

$$v_F=dc_F/dt \tag{6-15}$$

Based on eq. (6-11) to eq. (6-15), the rate of reactions can be rearranged as follows:

$$v=v_A/a=v_D/d=v_E/e=v_F/f \tag{6-16}$$

stoichiometric *adj.* 化学计量的

undesirability *n.* 不利条件

extent of reaction 反应进度

M6-34
单词读音及例句

homogeneous *adj.* 均相的

constant-volume 恒容

differential *n.* 微分

M6-35
单词读音及例句

gas-phase 气相

partial pressure 分压

M6-36
单词读音及例句

For homogeneous gas-phase reactions, it is often more convenient to express the rate law in terms of partial pressures.

3. Rate laws

The rate of reaction is often found to be proportional to the concentrations of the reactants raised to a power. For reaction with the form:

$$aA + dD \longrightarrow eE + fF$$

the rate of a reaction might be found to be proportional to the molar concentrations of two reactants A and D, so

$$v=kc^{a}_{A}c^{d}_{D} \tag{6-17}$$

proportionality *n.* 比例

formally *adv.* 正规地

M6-37
单词读音及例句

The constant of proportionality k is called the rate constant for the reaction. It is independent of the concentrations but depends on the temperature, solvent, catalyst and even the size and material of the reactor. The value of k reflects the reaction rates and how easy the reaction can proceed.

Notes:

It is independent of the concentrations but depends on the temperature, solvent, catalyst and even the size and material of the reactor.

译文：它不随物质的浓度变化，但和温度、溶剂、催化剂，甚至和反应器的大小、形状以及材质有关。

语法：本句为主系表结构，其中 is 为系动词，independent 为形容词作表语；be independent of 意为“独立于，不依赖；与……无关”。

An experimentally determined equation of this kind is called the rate law of the reaction. More formally, a rate law is an equation that expresses the rate of reaction in terms of the concentrations of all the species present in the overall chemical equation for the reaction at the time of interest:

$$v=f([A],[B], ...) \tag{6-18}$$

reflect *v.* 反映，反射

coincidence *n.* 巧合

underlying *adj.* 潜在的，基础的

M6-38
单词读音及例句

The rate law of a reaction is determined experimentally, and cannot in general be inferred from the chemical equation for the reaction. For example, the reaction of hydrogen and bromine has a very simple stoichiometry :

$$H_2(g) + Br_2(g) \longrightarrow 2\,HBr(g)$$

but its rate law is complicated:

$$v=\frac{k_a[H_2][Br_2]^{3/2}}{[Br_2]+k_b[HBr]} \tag{6-19}$$

In certain cases the rate law does reflect the stoichiometry of the reaction; but that is either a coincidence or reflects a feature of the underlying reaction mechanism.

4. Reaction order

Many reactions are found to have rate laws of the form

$$v=k_r[A]^a[B]^b \ldots \quad (6\text{-}20)$$

The power to which the concentration of a species (a product or a reactant) is raised in a rate law of this form is the order of the reaction with respect to that species. A reaction with the rate law in eq. (6-21) is first order in A and first order in B.

power *n.* （数）幂

with respect to 关于，谈到

first-order reaction 一级反应

$$v=k_r[A][B] \quad (6\text{-}21)$$

The overall order of a reaction with a rate law like that in eq. (6-20) is the sum of the individual orders, $a+b+\ldots$ The overall order of the rate law in eq. (6-21) is 1+1=2; the rate law is therefore said to be second-order overall.

M6-39
单词读音及例句

A reaction need not have an integral order, and many gas-phase reactions do not. For example, a reaction with the rate law as shown in eq. (6-22):

$$v=k_r[A]^{1/2}[B] \quad (6\text{-}22)$$

is half order in A, first order in B, and three-halves order overall.

Some reactions obey a zeroth-order rate law, and therefore have a rate that is independent of the concentration of the reactant. For example, the catalytic decomposition of phosphine (PH_3) on hot tungsten at high pressures has the rate law as follows:

zeroth-order 零级

phosphine *n.* 膦

tungsten *n.* 钨

participant *n.* 参与者

$$v=k_r \quad (6\text{-}23)$$

This law means that PH_3 decomposes at a constant rate until it has entirely disappeared.

When a rate law is not of the form in eq. (6-20), the reaction does not have an overall order and might not even have definite orders with respect to each participant.

M6-40
单词读音及例句

Reading comprehension

1. What does chemical kinetics deal with?
2. How to define the rates of reactions?
3. What is the overall order for a reaction with rate equation of $v=k_r[A]^2[B]$?

Exercise

1. True or false.

(1) The rate constant is independent of the concentrations of reactants.

(2) A reaction need not have an integral order.

(3) The rate law can be expressed in terms of partial pressures.

2. Translate the following sentences into English or Chinese.

(1) The rates of reactions are usually defined as the change of reactant and product concentration per unit time.______________________________

____

____

(2) 对于零级反应而言，反应速率与反应物浓度无关。__________

____

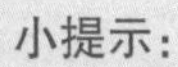

小提示：

• per unit time 译作“单位时间”

• be independent of... 与……无关

Further Reading

Monitoring the Reaction Progress

The first step in the kinetic analysis of reactions is to establish the stoichiometry of the reaction and identify any side reactions. The basic data of chemical kinetics are then the concentrations of the reactants and products at different times after a reaction has been initiated.

The rates of most chemical reactions are sensitive to the temperature, so in conventional experiments the temperature of the reaction mixture must be held constant throughout the course of the reaction. This requirement puts severe demands on the design of an experiment. Gas-phase reactions, for instance, are often carried out in a vessel held in contact with a substantial block of metal. Liquid-phase reactions must be carried out in an efficient thermostat.

Spectroscopy is widely applicable to the study of reaction kinetics, and is especially useful when one substance in the reaction mixture has a strong characteristic absorption in a conveniently accessible region of the electromagnetic spectrum. Commonly used methods of determining composition include emission spectroscopy, mass spectrometry, gas chromatography and nuclear magnetic resonance, which are shown in Figure 6-6. A reaction in which at least one component is a gas might result in an overall change in pressure in a system of constant volume, so its progress may be followed by recording the variation of pressure with time.

vessel *n.* 容器

substantial *adj.* 大量的

thermostat *n.* 恒温器

M6-41
单词读音及例句

Figure 6-6 Commonly used equipments of determing composition

mass spectroscopy 质谱
gas chromatography 气相色谱
nuclear magnetic resonance 核磁共振

M6-42
单词读音及例句

Reading comprehension

1. What is the major factor that affects the rate of a chemical reaction?

2. What are the commonly used spectroscopic methods for determining reaction rates?

Lesson five: Factors Affecting Reaction Rate

In this lesson you will learn:

- Main factors affecting rates
- Activation energy

aqueous *adj.* 水溶液的
interface *n.* 交界面

M6-43
单词读音及例句

1. Physical state

The physical state (solid, liquid, or gas) of a reactant is an important factor of the rate of reaction. When reactants are in the same phase, as in aqueous solution, thermal motion brings them into contact. However, when they are in separate phases, the reaction is limited to the interface between the reactants. Reactions can occur only at their area of contact, in the case of a liquid and a gas, at the surface of the liquid. Vigorous shaking and stirring may be needed to bring the reaction to completion. This means that the more finely divided a solid or liquid reactant, the greater its surface area per unit volume, and the more contact it makes with the other reactant, thus the faster the reaction.

Notes:

This means that the more finely divided a solid or liquid reactant, the greater its surface area per unit volume, and the more contact it makes with the other reactant, thus the faster the reaction.

译文：这意味着固体或液体反应物颗粒越精细，单位体积的表面积越大，与其他反应物的接触越多，则反应越快。

语法：句中包含一连串的“the+ 比较级，the+ 比较级”的结构，意为“越……，越……”。

crush *v.* 挤压
collision *n.* 碰撞
malic acid 苹果酸

M6-44
单词读音及例句

2. Surface area of solid state

In a solid, only those particles that are at the surface can be involved in a reaction. Crushing a solid into smaller parts means that more particles are present at the surface, and the frequency of collisions between these and reactant particles increases, and so reaction occurs more rapidly. For example, Sherbet (powder) is a mixture of very fine powder of malic acid (a weak organic acid) and sodium hydrogen carbonate. On contact with the saliva in the mouth, these chemicals quickly dissolve and react, releasing carbon dioxide and providing for the fizzy sensation. Also, finely divided aluminium confined in a shell explodes violently. If larger pieces of aluminium are used, the reaction is much slower and only sparks are seen.

saliva *n.* 唾液，口水
fizzy *adj.* 有气的，嘶嘶的
spark *n.* 火星，火花

M6-45
单词读音及例句

3. Concentration and pressure

The reactions are due to collisions of reactant species. The frequency with which the molecules or ions collide depends upon their concentrations. The more crowded the molecules are, the more likely they are to collide and react with one another. Thus, an increase in the concentrations of the reactants will usually result in the corresponding increase in the reaction rate, while a decrease in the concentrations will usually have a reverse effect. For example, combustion will occur more rapidly in pure oxygen than in air which contains 21% oxygen.

Sherbet powder

The rate equation shows the detailed dependence of the reaction rate on the concentrations of reactants and other species present. The mathematical forms depend on the reaction mechanism. The actual rate equation for a given reaction is determined experimentally and provides information about the reaction mechanism.

Increasing the pressure in a gaseous reaction will increase the number of collisions between reactants, increasing the rate of reaction. This is because the activity of a gas is directly proportional to the partial pressure of the gas. This is similar to the effect of increasing the concentration of a solution.

pre-exponential factor
指前因子，频率因子

acivation energy
活化能

molar gas constant
气体常数

4. Temperature

Temperature usually has a major effect on the rate of a chemical reaction. Molecules at a higher temperature have more thermal energy. Although collision frequency is greater at higher temperatures, this alone contributes only a very small proportion to the increase in rate of reaction. Much more important is the fact that the proportion of reactant molecules with sufficient energy to react (energy greater than activation energy: $E > E_a$) is significantly higher.

The effect of temperature on the reaction rate constant usually obeys the Arrhenius equation:

$$k=Ae^{-E_a/(RT)} \tag{6-24}$$

M6-46
单词读音及例句

where A is the pre-exponential factor or A-factor, E_a is the activation energy, R is the molar gas constant and T is the absolute temperature.

At a given temperature, the chemical rate of a reaction depends on the value of the A-factor, the magnitude of the activation energy, and the concentrations of the reactants. Usually, rapid reactions require relatively small activation energies.

magnitude *n.*
级数，广大

rule of thumb
经验法则

misconception *n.*
误解，错觉

The "rule of thumb" that the rate of chemical reactions doubles for every 10 ℃ temperature rise is a common misconception. This may have been generalized from the special case of biological systems. The kinetics of rapid reactions can be studied with the temperature jump method. This involves using a sharp rise in temperature and observing the relaxation time of the return to equilibrium. A particularly useful form of temperature jump apparatus is a shock tube, which can rapidly increase a gas's temperature by more than 1000 degrees.

M6-47
单词读音及例句

5. Catalysts

A catalyst is a substance that alters the rate of a chemical reaction but it remains chemically unchanged afterwards. The catalyst increases the rate of the reaction by providing a new reaction mechanism to occur within a lower activation energy. A catalyst does not affect the position of the equilibrium, as the catalyst speeds up the backward and forward reactions equally.

hypothetical *adj.*
假设的，假定的

pathway *n.*
路径，通道

Figure 6-7 shows the effect of a catalyst in a hypothetical endothermic chemical reaction. The presence of the catalyst opens a new reaction pathway (shown in blue) with a lower activation energy. The final result and the overall thermodynamics are the same.

M6-48
单词读音及例句

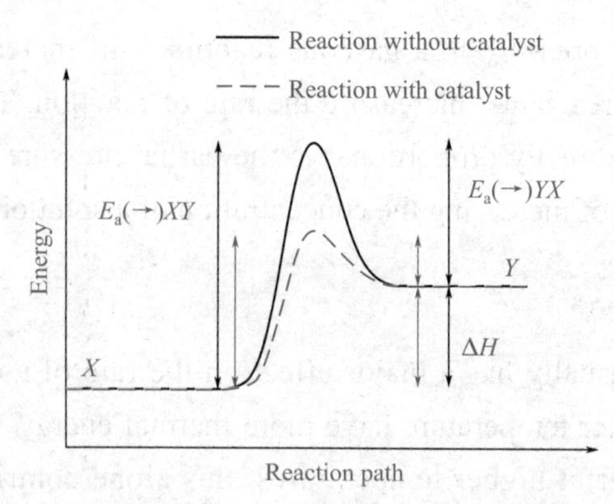

Figure 6-7 Generic potential energy diagram

Reading comprehension

1. How to increase the reaction rate for solid state reactants?
2. What do higher concentrations of the reactants result in?
3. How does a catalyst increase the rate of a chemical reaction?

Exercise

1. True or false.

(1) In a solid, only those particles that are at the surface can be involved in a reaction.

(2) Molecules at a higher temperature are likely to have sufficient energy to react.

(3) The catalyst increases the rate of the reaction by lowering the position of the equilibrium.

2. Translate the following sentences into English or Chinese.

(1) The "rule of thumb" that the rate of chemical reactions doubles for every 10 ℃ temperature rise is a common misconception.________________

(2) 温度对反应速率常数的影响通常遵循阿伦尼乌斯方程。__________

小提示：

- every...
  译作：每……
- "rule of thumb"
  译作：经验法则
- "……对……的影响"英文表达：effect of...on...

Activation Energy and the Arrhenius Law

Imagine waking up on a day when you have lots of fun stuff planned. Does it ever happen that, despite the exciting day that lies ahead, you need to muster some extra energy to get yourself out of bed? Once you're up, you can coast through the rest of the day, but there's a little hump you have to get over to reach that point.

The activation energy of a chemical reaction is kind of like that "hump" you have to get over to get yourself out of bed. Even energy-releasing (exergonic) reactions require some amount of energy input to get going, before they can proceed with their energy-releasing steps.

Activation energy is denoted by E_a and typically has units of kilojoules per mole. The term "activation energy" was introduced by the Swedish scientist Svante Arrhenius in 1889. The Arrhenius equation relates activation energy to the rate at which a chemical reaction proceeds:

$$k=A\exp[-E_a/(RT)] \tag{6-25}$$

where k is the reaction rate coefficient, A is the frequency factor for the reaction, E_a is the activation energy, R is the universal gas constant, and T is the absolute temperature (Kelvin).

Yet not only chemists but many scientists from other disciplines do apply a simple Arrhenius law to their results and thus obtain the so-called "effective" value of E_a. Then, basing their discussion on its value, they can speculate about a mechanism for the process involved or simply extrapolate the data to new conditions. In this connection it is necessary to mention some "unconventional applications of the Arrhenius law". Scientists have demonstrated how it could be applied to such phenomena as chirping of tree crickets (E_a=51 kJ/mol), creeping of ants (108 kJ/mol below 16 ℃ and 51 kJ/mol above that temperature), flashing of fireflies (51 kJ/mol), terrapin heart-beating (77 kJ/mol from 18 to 34 ℃ and much higher below 18 ℃), emitting of human alpha brain-wave (about 30 kJ/mol), and even counting and forgetting of human brain (100 kJ/mol for both). J. A. Campbell, the author of a remarkable popular book on chemical kinetics, presented other examples in his Eco-chem column in this journal. The heartbeat rate of the water flea Daphnia increases with the temperature (E_a=50 kJ/mol). A rapid change in the rate of bacterial hydrolysis of fish muscle as T varies near 0 ℃ yields E_a=1420 kJ/mol! But this value

despite　*adv.* 尽管，虽然

hump　*n.* 隆起，小峰

denote　*v.* 表明，指代

M6-49
单词读音及例句

not only...but also... 不但……而且……

extrapolate　*v.* 推算，推断

unconventional　*adj.* 非常规的

chirp　*vi.* 鸟叫，虫鸣

creep　*v.* 爬行，蹑手蹑脚地移动

span　*v.* 持续，横跨，涵盖

M6-50
单词读音及例句

is very uncertain because of a very narrow temperature range (3.3 ℃). Perhaps even more important is the fact that the temperature range spans the water freezing point; so the rate of a bacterial process can change dramatically.

Reading comprehension

1. How to describe activation energy?
2. What does the Arrhenius law state?

项目三 化工基础知识

Part Ⅲ: Basis of Chemical Engineering

第七单元　反应器

Unit 7　Reactor

Lesson one: The Stirred Tank Reactor

In this lesson you will learn:

- Chemical reactors
- The stirred tank reactor
- The continuous stirred tank reactor

The chemical reactor is the heart of any chemical process. Chemical processes turn inexpensive chemicals into valuable ones, and chemical engineers are the only people technically trained to understand and handle them. While separation units are usually the largest components of a chemical process, their purpose is to purify raw materials before they enter the chemical reactor and to purify products after they leave the reactor. Here is a very generic flow diagram of a chemical process as shown in Figure 7-1.

separation units 分离装置

purify *v.* 净化，使纯净

byproduct *n.* 副产品

M7-1
单词读音及例句

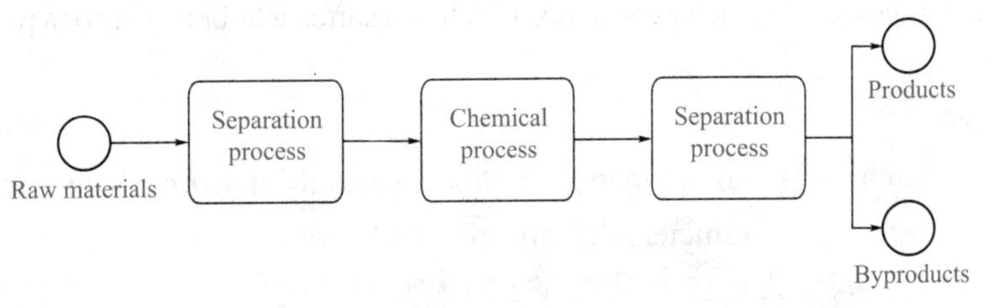

Figure 7-1　A generic flow diagram of a chemical process

externally　*adv.*
外部地

impure　*adj.*
不纯的

dominate　*v.*
控制，在……中占主要地位

M7-2
单词读音及例句

Raw materials from another chemical process or purchased externally must usually be purified to a suitable composition for the reactor to handle. After leaving the reactor, the unconverted reactants, any solvents, and all byproducts must be separated from the desired product before it is sold or used as a reactant in another chemical process.

The key component in any process is the chemical reactor; if it can handle impure raw materials or not produce impurities in the product, the savings in a process can be far greater than if we simply build better separation units. In typical chemical processes the capital and operating costs of the reactor may be only 10% to 25% of the total, with separation units dominating the size and cost of the process. Yet the performance of the chemical reactor totally controls the costs and modes of operation of these expensive separation units, and thus the chemical reactor largely controls the overall economics of most processes. Improvements in the reactor usually have enormous impact on upstream and downstream separation processes. (From Lanny D.Schmidt, THE ENGINEERING OF CHEMICAL REACTIONS, 2E, 2005, 618.)

1. The batch stirred tank reactor

turbine　*n.* 涡轮

propeller　*adj.*
螺旋桨

spatially　*adv.*
空间地

uniform　*adj.*
均衡的，一致的

M7-3
单词读音及例句

A batch reactor is defined as a closed spatially uniform system which has concentration parameters that are specified at time zero. This requires that the system either be stirred rapidly or started out spatially uniform so that stirring is not necessary. A batch stirred tank reactor might look as illustrated in Figure 7-2.

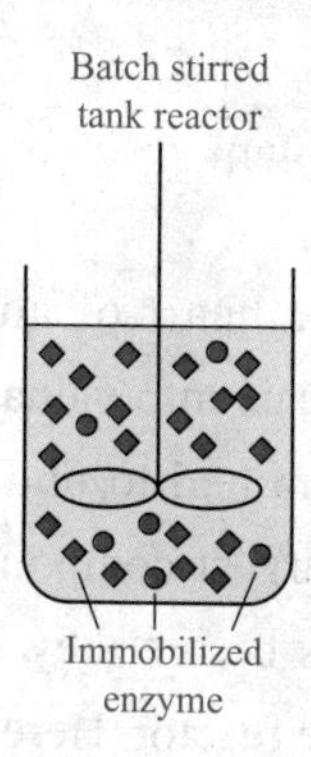

Figure 7-2　Schematic of batch stirred tank reactor

It is composed of a reactor and a mixer such as a stirrer, a turbine wing or a propeller.

Notes:

A batch reactor is defined as a closed spatially uniform system which has concentration parameters that are specified at time zero.

译文：间歇式反应器是指在零时刻具有特定（物料）浓度的封闭均一系统。

语法：句中 which 为关系代词，引导定语从句，指代 system，在此定语从句中，又包含一个关系代词 that 引导的定语从句，that 指代 concentration parameters。

Composition and temperature are therefore independent of position in thc reactor, so that the number of moles of species j in the system N_j is a function of time alone. Since the system is closed (no flow in or out), we can write simply that the change in the total number of moles of species j in the reactor is equal to the stoichiometric coefficient v_j multiplied by the rate multiplied by the volume of the reactor

$$\frac{dN_j}{dt}=Vv_j r$$

for a single reaction.

The number of moles of species j in a batch reactor is simply the reactor volume V times the concentration

$$N_j=VC_j$$

so the above equation becomes

$$\frac{d(VC_j)}{dt}=V\frac{dC_j}{dt}+C_j\frac{dV}{dt}=Vv_j r(C_j)$$

If the reactor is at constant volume, then we can divide each term by V to yield

$$\frac{d(C_j)}{dt}=v_j r(C_j)$$

which is only valid if the volume of the reactor does not change.

species *n.* 物料，种类

moles *n.* 摩尔（复数）

stoichiometric *adj.* 化学计量的

coefficient *n.* 系数，率

M7-4
单词读音及例句

2. The continuous stirred tank reactor

Reactors can be operated in batch (no mass flow into or out of the reactor) or flow modes. Flow reactors operate between limits of completely unmixed contents (the plug-flow tubular reactor or PFTR) and completely mixed contents (the continuous stirred tank reactor or CSTR). A flow reactor may be operated in steady state (no variables vary with time) or transient modes.

Here we consider the situation where mixing of fluids is sufficiently rapid that the composition does not vary with position in the reactor. This is a continuous stirredtank reactor or CSTR as shown in Figure 7-3.

Our picture is that of a tank with a stirring propeller that is fed and drained by pipes containing reactants and products, respectively. In this situation the crucial feature is that the composition is identical everywhere in the reactor and in the exit pipe. Nothing is a function of position except between the inlet pipe and the reactor entrance, where mixing is assumed to occur instantly!

continuous *adj.* 连续的，持续的

transient *adj.* 短暂的，过渡的

fed *v.*（feed 的过去式）给机器加料

drain *v.* & *n.* 排水，出料，排水管

pipe *n.* 管道

M7-5
单词读音及例句

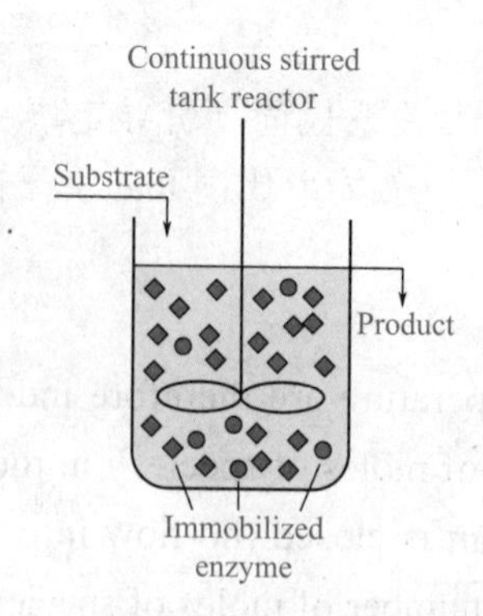

Figure 7-3　Schematic of continuous stirred tank reactor

> Notes:
>
> Nothing is a function of position except between the inlet pipe and the reactor entrance, where mixing is assumed to occur instantly!
>
> 译文：假定（反应器）中混合过程是瞬间发生的，除了在入口管和反应器入口之间，没有任何关于位置的函数。
>
> 语法：句中 where 为关系副词，引导非限制性定语从句。

integral　*adj.* 整体的，完整的
inlet　*n.* 入口
outlet　*v.* 出口
dimension　*n.* 尺寸，量纲
volumetric flow 体积流量

M7-6
单词读音及例句

The idea that the composition is identical everywhere in the reactor and in the exit pipe requires some thought. It might seem that, since the concentration changes instantly at the entrance where mixing occurs, reaction occurs there and nothing else happens in the reactor because nothing is changing. However, reaction occurs throughout the reactor, but mixing is so rapid that nothing appears to change with time or position.

Since the reactor is assumed to be uniform in composition everywhere, we can make an integral mass balance on the number of moles N_j of species j in a reactor of volume V. This gives

$$\frac{\mathrm{d}N_j}{\mathrm{d}t} = F_{j0} - F_j + V\nu_j r$$

where F_{j0} and F_j are molar flow rates of species j (in moles/time) in the inlet and outlet, respectively. Each term in this equation has dimensions of moles/time. This equation as written is exact as long as the reactor is completely mixed.

We can relate the molar flow rates F_{j0} and F_j of species j to the concentration by the relationships

$$F_{j0}=v_0C_{j0}$$

and

$$F_j=vC_j$$

respectively, where v_0 and v are the volumetric flow rates into and out of the reactor.

If the density of the fluid is constant, then the volumetric flow rates in and out of the reactor are equal.

$$v = v_0$$

If we assume that V is constant and the density does not change with composition, the mass-balance equation then simplifies to become

$$V\frac{\mathrm{d}(C_j)}{\mathrm{d}t} = v(C_{j0} - C_j) + Vv_j r$$

Next we assume that compositions are independent of time (steady state) and set the time derivative equal to zero to obtain

$$v(C_{j0} - C_j) + Vv_j r = 0$$

We call the volume divided by the volumetric flow rate the *reactor residence time*

$$\tau = \frac{V}{v}$$

With these approximations we write the steady-state mass balance on species j in the CSTR as

$$C_{j0} - C_j = -\tau v_j r$$

This will be the most used form of the mass-balance equation in the CSTR.

derivative *n.* 派生物，导数

residence time 停留时间

approximation *n.* 近似值

M7-7
单词读音及例句

Reading comprehension

1. What is the key component in chemical process?

2. Is stirring necessary for a batch reactor?

3. Why is the composition identical everywhere in the reactor and in the exit pipe?

Exercise

Translate the following sentences into Chinese.

（1）The chemical reactor is the “unit” in which chemical reactions occur.____

____

____

（2）This equation as written is exact as long as the volume of the reactor does not change.________________________________

____

Further Reading

Application of Batch Reactors

Most industrial reactors are operated in a continuous mode instead of batch because continuous reactors produce more product with smaller equipment, require less labor and maintenance, and frequently produce better quality control. Continuous processes are more difficult to start and stop than batch reactors, but they make product without stopping to change batches and they require minimum labor. Batch processes can be tailored to produce small amounts of product when needed. Batch processes are also ideal to measure rates and kinetics in order to design continuous processes: Here one only wants to obtain information rapidly without generating too much product that must be disposed of. In pharmaceuticals batch processes are sometimes desired to assure quality control: Each batch can be analyzed and certified (or discarded), while contamination in a continuous process will invariably lead to a lot of worthless product before certifiable purity is restored. Food and beverage are still made in batch processes in many situations because biological reactions are never exactly reproducible, and a batch process is easier to "tune" slightly to optimize each batch. Besides, it is more romantic to produce beer by "beechwood aging", wine by stamping on grapes with bare feet, steaks by charcoal grilling, and similar batch processes.

We will develop mass balances in terms of mixing in the reactor. In one limit the reactor is stirred sufficiently to mix the fluid completely, and in the other limit the fluid is completely unmixed. In any other situation the fluid is partially mixed, and one cannot specify the composition without a detailed description of the fluid mechanics.

Reading comprehension

1. Should beer be made in batch or continuous processes?
2. Describe the benefits of batch processes.

Lesson two: Tubular Reactor

In this lesson you will learn:

- Tubular reactor
- The plug-flow tubular reactor
- Jacketed PFTR

Tubular reactors are widely used in industry, particularly with exothermic gas-phase reactions that require a solid catalyst. These tubular reactors (Figure 7-4) are often operated adiabatically, which yields a monotonically increasing temperature profile with the maximum temperature occurring at the exit. There is a tradeoff between reactor size and recycle flow rate. If high heat-transfer rates are required, small-diameter tubes are used to increase the surface area to volume ratio. Several tubes may be arranged in parallel, connected to a manifold or fitted into a tube sheet in a similar arrangement to a shell and tube heat exchanger. For high-temperature reactions the tubes may be arranged in a furnace.

exothermic　*adj.* 放热的

phase　*n.* 相，阶段

tubular　*adj.* 管状的

adiabatically　*adv.* 绝热地

monotonically　adv. 单调地

in parallel　并联的，并行的

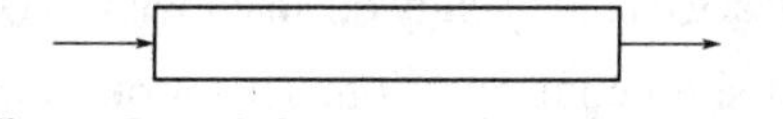

Figure 7-4　Schematic of tubular reactor

1. The plug-flow tubular reactor

The CSTR is completely mixed. The other limit where the fluid flow is simple is the tubular reactor, where the fluid is completely unmixed and flows down the tube as a plug.

M7-8
单词读音及例句

Notes:

The other limit where the fluid flow is simple is the tubular reactor, where the fluid is completely unmixed and flows down the tube as a plug.

译文：另一种流体流动简单的极限是管式反应器，其中流体完全未混合，并像塞子一样沿管流下。

语法：句中第一个 where 为关系副词，引导定语从句，第二个 where 也是关系副词，引导非限制性定语从句。

Here we picture a pipe through which fluid flows without dispersion and maintains a constant velocity profile, although the actual geometry for the plug-flow approximation may be much more complicated. Simple consideration shows that this situation can never exist exactly because at low flow rates the flow profile will be laminar, while at high flow rates turbulence in the tube causes considerable axial mixing. Nevertheless, this is the limiting case of no mixing, and the simplicity of solutions in the limit of perfect plug-flow makes it a very useful model.

plug　*n.* 塞子

laminar　*adj.* 层流的

turbulence　*n.* 湍流

axial　*adj.* 轴向的

M7-9
单词读音及例句

We must develop a differential mass balance of composition *versus* position and then solve the resulting differential equation for $C_j(z)$ and $C_j(L)$ (Figure 7-5). We consider a tube of length with position z going from 0 to L. The molar flow rate of species j is F_{j0} at the inlet (z=0), $F_j(z)$ at position z and $F_j(L)$ at the exit L. The mass balance on species j is

$$u\frac{\mathrm{d}C_j}{\mathrm{d}z} = v_j r$$

which is the form of the PFTR equation we will most often use.

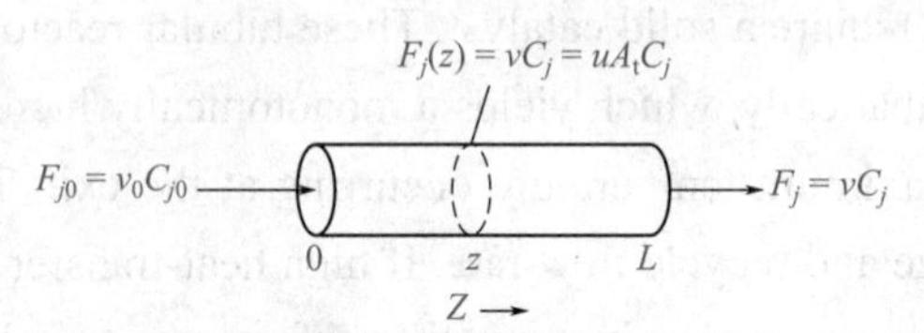

Figure 7-5 Schematic of plug-flow

2. Jacketed PFTR

The cooled tubular reactor is frequently operated with a cooling jacket surrounding the reactor in a tube-and-shell configuration. This looks simply like a tube-in-shell heat exchanger, typically with catalyst in the tube to catalyze the reaction. A major example of this reactor is ethylene oxidation to ethylene oxide (EO)

$$C_2H_4 + \frac{1}{2}O_2 \longrightarrow C_2H_4O$$

$$C_2H_4 + 3O_2 \longrightarrow 2CO_2 + 2H_2O$$

The heat release in the second reaction is much greater, so that much more heat is released if the selectivity to EO decreases. For world-scale ethylene oxide plants the reaction is run in a reactor consisting of several thousand 1-in-diameter-20-ft-long (1in=0.0254 m, 1ft=0.3048 m) tubes in a tube-and-shell heat exchanger. Heat transfer is accomplished using boiling recirculating hydrocarbon on the shell side to absorb the reaction heat. This heat transfer method assures that the temperature of the tubes is constant at the boiling point of the fluid.

However, with finite flow rate in the jacket, the coolant may heat up, and we must now distinguish between cocurrent and countercurrent flows. While it is true that countercurrent operation gives better heat transfer for an ordinary exchanger, this is not usually desired for a tubular reactor. Heat is generated mostly near the reactor entrance, and the maximum amount of cooling is needed there. In countercurrent flow, the coolant has been heated before it reaches this region, while in cocurrent operation, the coolant is the coldest exactly where the heat load is greatest. Thus cocurrent operation reduces the hot spot, which can plague cooled tubular reactors with exothermic reactions.

cooling jacket 冷却套管

ethylene *n.* 乙烯

oxidation *n.* 氧化作用

ethylene oxide 环氧乙烷

recirculating 再循环的

hydrocarbon *n.* 烃

M7-10 单词读音及例句

countercurrent *n.* 逆流

cocurrent *n.* 并流

coolant *n.* 冷却剂

hot spot 热点

plague *v.* & *n.* 使折磨，使苦恼，瘟疫

M7-11 单词读音及例句

Notes:

In countercurrent flow, the coolant has been heated before it reaches this region, while in cocurrent operation, the coolant is the coldest exactly where the heat load is greatest.

译文：当逆流流动时，冷却剂在达到此区域（进口处）之前已经被加热了；而并流流动时，冷却剂在热负荷最大时恰好是最冷的。

语法：句中连词 while 引导状语从句，表示对比，意思是“然而，而”。

A major goal in wall cooling is to spread out the hot zone and prevent very high peak temperatures. High peak temperatures cause poor reaction selectivity, cause carbon formation, deactivate catalysts, and cause corrosion problems in the reactor walls. Cocurrent flows spread out the hot zone and cause lower peak temperatures, but many additional design features must be considered in designing jacketed reactors.

Reading comprehension

1. What reaction processes are tubular reactors suitable for?

2. Describe the characteristics of PFTR.

3. Which one has better heat transfer for the cooled tubular reactors: cocurrent or countercurrent flows?

Exercise

1. True or false.

（1）The shape of a tubular reactor must be a long tube.

（2）PFTR is the limiting case of no mixing.

（3）A jacketed PFTR always looks simply like a tube-in-shell heat exchanger.

2. Translate the following sentences into Chinese.

（1）The equipment used for these reactors in fact looks very similar to heat exchangers.____

小提示：
• 与……相似：similar to

（2）The cost of several smaller reactors is usually greater than one large reactor.

Further Reading

Design of Cooled Tubular Reactor

As reactor inlet temperature is lowered, smaller recycle flow rates can be used for a given maximum-allowable reactor exit temperature, but reactor size increases. The geometry of the reactor (tube diameter, tube length, and number of tubes) can be easily adjusted to make sure whatever pressure drop over the reactor is dictated by the economics of compression costs. For example, for a specified pressure drop and total flow rate through the reactor, the reactor length is fixed at a reasonable value (perhaps 3-6 m) as determined by physical and mechanical limitations, and reactor diameter is calculated from the given pressure drop. Since no heat transfer occurs, tube diameter is not an issue.

internal *adj.* 内部的

intermediate *adj.* 中间的，过渡的

M7-12
单词读音及例句

However, it is often more economical to design a tubular reactor with internal cooling. This permits smaller recycle flow rates for the same maximum temperature limitation. Now the peak temperature occurs not at the exit of the reactor but at some intermediate axial location. The standard cooled reactor configuration consists of multiple tubes packed with catalyst. Process gas flows through these tubes, and heat is transferred through the tube wall to a coolant on the shell side of the vessel. For high-temperature reactions, steam is often generated on the shell side to remove the exothermic heat of reaction.

Heat transfer is a critical part of these cooled reactors, so tube diameter becomes an important design optimization parameter. The smaller the tube diameter, the larger the heat-transfer area. However, the smaller the diameter, the larger the pressure drop because of the higher velocity. The economic advantage of designing for small pressure drop is obvious: it reduces compression costs.

The advantage of designing for large heat-transfer area is less obvious. Typical superficial velocities are low (0.5 m/s), so overall heat-transfer coefficients that occur in packed gas phase tubular reactors are quite small. The large heat-transfer area associated with small-diameter tubes means small temperature differentials between the hot reactor temperature and the coolant temperature. Since peak temperature is typically limited to some maximum value to prevent catalyst degradation or side reactions, large tube diameters require low coolant temperatures. For the same peak temperature, a small-diameter tube gives higher average temperature down the length of the reactor. This means the reactor can be smaller (requires less catalyst). Thus small tubes reduce reactor cost but increase compression cost. In addition, large design temperature differentials can lead to poor dynamic controllability. This occurs because it is more difficult to change heat-transfer rates (by changing temperature differentials) when disturbances upset the reactor. (From Ind.Eng.Chem.Res.,Vol.40,No.24,2001)

Reading comprehension

1. How to calculate the reactor diameter for a specified pressure drop and total flow rate through the reactor?

2. What is the relationship between tubes diameter and heat transfer area?

Lesson three: Fluidized and Fixed Bed Reactors

In this lesson you will learn:

- Catalytic reactions
- Fixed bed reactors
- Fluidized bed reactors

1. Catalytic reactions

Catalysts are substances added to a chemical process that do not enter into the stoichiometry of the reaction but that cause the reaction to proceed faster or make one reaction proceed faster than others.

We distinguish between homogeneous and heterogeneous catalysts. Homogeneous catalysts are molecules in the same phase as the reactants (usually a liquid solution), and heterogeneous catalysts in another phase (usually solids whose surfaces catalyze the desired reaction). Acids, bases, and organometallic complexes are examples of homogeneous catalysts, while solid powders, pellets, and reactor walls are examples of heterogeneous catalysts. By far the most efficient catalysts are enzymes, which regulate most biological reactions. Enzymes are proteins that may be either isolated molecules in solution (homogeneous) or molecules bound to large macromolecules or to a cell wall (heterogeneous).

catalyst *n.* 催化剂

stoichiometry *n.* 化学计量学

organometallic *adj.* 有机金属的

pellet *n.* 小球

M7-13
单词读音及例句

Notes:

Enzymes are proteins that may be either isolated molecules in solution (homogeneous) or molecules bound to large macromolecules or to a cell wall (heterogeneous).

译文：酶是蛋白质，可以是溶液中的孤立分子（均相），也可以是与生物大分子或细胞壁结合的分子（非均相）。

语法：句中 that 为关系代词，引导定语从句，指代 proteins。

For simple reactions, the effect of the presence of a catalyst is to

（1）increase the rate of reaction,

（2）permit the reaction to occur at a lower temperature,

（3）permit the reaction to occur at a more favorable pressure,

（4）reduce the reactor volume,

（5）increase the yield of a product(s).

Catalytic or heterogeneous reactors are an alternative to homogenous reactors. If a solid catalyst is added to the reactor, the reaction is said to be heterogeneous. The first complication with catalytic processes is that we need to maintain the catalyst in the reactor. With a homogeneous catalyst (catalyst and reactants in the same phase), the only method to reuse the catalyst is to separate it from the products after the reactor and recycle the catalyst into the feed streams. Much more preferable and economical is to use a heterogeneous catalyst that is kept inside the reactor such as in a powder that is filtered from the product or as pellets packed in the reactor. (From Lanny D.Schmidt, THE ENGINEERING OF CHEMICAL REACTIONS, 2E, 2005, 618.)

cylindrical *adj.* 圆柱形的

spherical *adj.* 圆球形的

lining *n.* 衬里，内层

distributor *n.* 分布器

steel alloy 钢合金

thermocouple *n.* 热电偶温度计

thermometers *n.* 温度计

2. The packed bed reactor

The most used industrial reactor is the catalytic packed bed reactor. The shape of a fixed bed catalytic reactor is either cylindrical, or occasionally spherical and the flow throughout the bed may be upward or downward. Most of these reactors are designed with an internal lining and a distributor to control the gas passage through the catalyst bed. The internal reactor volume is occupied by the catalyst and any inert fill above and/or below the bed. The shell of the reactor is usually made of either steel alloy or carbon steel which is resistant to hydrogen and/or hydrogen sulfide. Steel alloy is employed for reactors that are in hydrogen and hydrogen sulfide service. It is not uncommon for these units to operate at or near 1000 ℉; temperature measurements are normally obtained with thermocouples rather than thermometers.

M7-14
单词读音及例句

Notes:

It is not uncommon for these units to operate at or near 1000 ℉; temperature measurements are normally obtained with thermocouples rather than thermometers.

译文：这些设备的操作温度达到1000 ℉不稀奇，通常采用热电偶温度计，而非双金属温度计测量温度。

语法：为防止句子结构头重脚轻，使用句首的it为先行的形式主语，代表后移的不定式to operate at or near 1000 ℉。

A fixed bed reactor contains small catalyst particles that are approximately 0.1 to 0.2 inches in diameter through which the reaction mixture is passed. Fill material is often used for support and to moderate temperature changes. The packing itself may be arranged in any of several different ways (Figure 7-6):

(a) A single large bed.

(b) Several parallel packed tubes in a single shell.

(c) Several beds, each in its own shell.

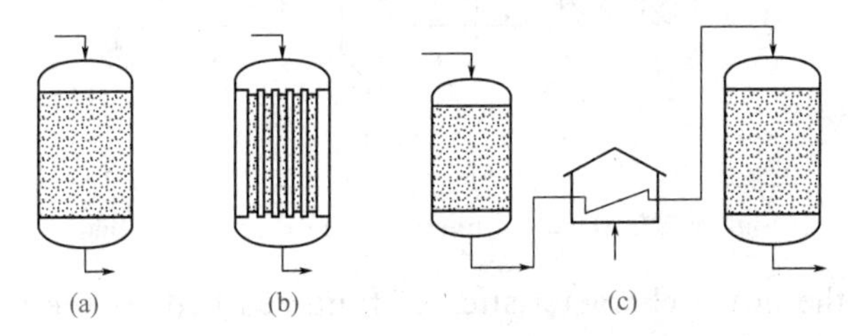

Figure 7-6 Fixed bed reactors, (a) single bed; (b) parallel packed tubes; (c) several beds connected in series

Fixed catalytic bed reactors have significant advantages relative to other types of heterogeneous catalytic reactors (particularly the fluid bed unit). (From Louis Theodore, Chemical reactor analysis and application for the practicing engineer, 2012, 577)

3. The fluidized bed reactor

The other major type of catalytic reactor is a situation where the fluid and the catalyst are stirred instead of having the catalyst fixed in a bed. If the fluid is a gas, it is difficult to stir solid particles and gas mechanically, but this can be simply accomplished by using very small particles and flowing the gas such that the particles are lifted and gas and particles swirl around the reactor. We call this reactor a fluidized bed.

The solid substrate (the catalytic material upon which chemical species react) material in the fluidized bed reactor is typically supported by a porous plate, known as a distributor. The fluid is then forced through the distributor up to the

particle *n.* 颗粒

swirl *v.* 盘绕，打旋

porous *adj.* 多孔的，有气孔的

M7-15
单词读音及例句

incipient　*adj.* 初期的，初始的

flow regime 流动区

M7-16 单词读音及例句

solid material (Figure 7-7). At lower fluid velocities, the solids remain in place as the fluid passes through the voids in the material. This is known as a packed bed reactor. As the fluid velocity is increased, the reactor will reach a stage where the force of the fluid on the solids is enough to balance the weight of the solid material. This stage is known as incipient fluidization and occurs at this minimum fluidization velocity. Once this minimum velocity is surpassed, the contents of the reactor bed begin to expand and swirl around much like an agitated tank or boiling pot of water. The reactor is now a fluidized bed. Depending on the operating conditions and properties of solid phase various flow regimes can be observed in this reactor.

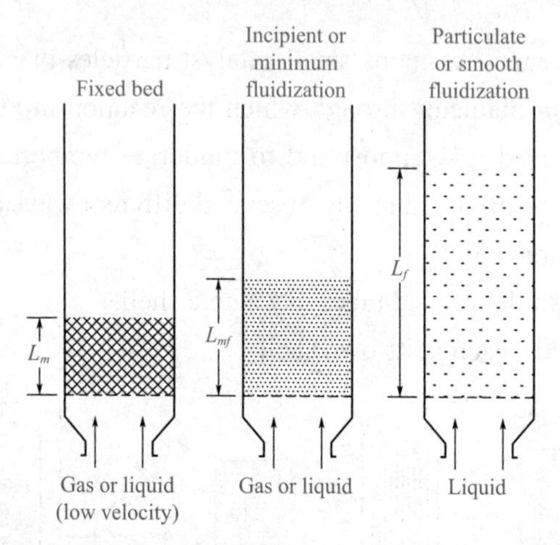

Figure 7-7　Types of particle-fluid contact in a bed

oil cracking 石油裂解

zinc coating　镀锌

desulfurization *n.* 脱硫作用

granulation *n.* 造粒

M7-17 单词读音及例句

One of the novel characteristics of fluidized beds is the uniformity of temperature found throughout the system. Essentially constant conditions are known to exist in both the horizontal and vertical directions in both short and long beds. This homogeneity is due to the turbulent motion and rapid circulation rate of the solid catalyst particles within the fluid stream described above. Temperature variations can occur in some beds in regions where quantities of relatively hot or cold catalyst particles are present but these effects can generally be neglected. Consequently, fluidized beds find wide application in industry, *e.g.*, oil cracking, zinc coating, coal combustion, gas desulfurization, heat exchangers, plastics cooling, and fine powder granulation.

Notes:

Essentially constant conditions are known to exist in both the horizontal and vertical directions in both short and long beds.

译文：已知在短层和长层中的水平和垂直方向都存在基本恒等的条件。

语法：本句采用被动语态，句中含有两个“both... and...”并列结构，意为“……和……（两者）都”。

Reading comprehension

1. What is the effect of adding a catalyst to the reaction?

2. What is the difference between a fixed bed reactor and a fluidized bed reactor?

3. Describe the minimum fluidization speed.

Exercise

1. True or false.

（1）Biological enzyme is a highly efficient catalyst, which could catalyze the reaction under mild conditions.

（2）Thermometers are more suitable for high temperature measurement than thermocouple thermometers.

（3）The fluid flowing in the fluidized bed is gas or liquid, which does not contain solid particles.

2. Translate the following sentences into English or Chinese.

（1）Biological catalysts are without question the most important catalysts (to us) because without them life would be impossible.______________________

____

（2）固定床和流化床反应器都是工业中常用的反应器。______________

____

小提示：

• both ... and... 两者都是

Further Reading

Fluidized Bed Membrane Reactors

Fluidized bed membrane reactors (FBMRs), also called membrane assisted fluidized bed reactors, are a special type of reactor in which the permeable membranes are integrated inside a fluidized reaction bed to intensify the reaction process. Such integration not only provides the advantages of both the fluidized bed and the membrane reactor, but also accomplishes a synergistic effect.

The main advantages of FBMRs include:

• Isothermal operation. The fluidized bed reactor has excellent tube-to-bed heat transfer properties, allowing a safe and efficient reactor operation even for highly exothermic reactions. Also, for highly endothermic reactions, where the hot catalyst is circulated between the reactor and the regenerator, the excellent gas-to-solid heat transfer

permeable *adj.* 能透过的，有渗透性的

synergistic *adj.* 协同的，协同作用的

isothermal *adj.* 等温的，等温线的

exothermic *adj.* 放热的

endothermic *adj.* 吸热的

M7-18
单词读音及例句

characteristics of the fluidized beds can be exploited effectively. The intense macroscale solids mixing induced by the rising bubbles results in a remarkable temperature uniformity.

• Negligible pressure drop, and no internal mass and heat transfer limitations because of the small particle sizes of the catalysts employed.

• Flexibility in membrane heat transfer surface area and arrangement of the membranes. Optimal concentration profiles can be created *via* distributive feeding of one of the reactants or selective withdrawal of one of the products.

• Improved fluidization behavior due to the presence of the inserts and gas permeation through the membranes, leading to reduced axial gas back-mixing and enhanced bubble breakage, while the bubble-to-emulsion mass transfer and membrane permeation are also improved due to the reduced bubble size.

Therefore, large improvements in conversion and selectivity may be achieved in FBMRs. Moreover, the membrane area can be reduced remarkably compared with the packed bed membrane reactors.

However, there are some disadvantages of FBMRs, including:

• Difficulties in reactor construction and membrane sealing at the wall.

• Erosion of reactor internals and catalyst attrition.

Nevertheless, the FBMR has become the most recent trend to overcome the limitations often prevailing in packed bed membrane reactors.

distributive *adj.* 分布的，分配的

withdrawal *n.* 收回，采出

back-mixing 返混

emulsion *n.* 乳剂，乳状液

sealing *vt.* & *n.* 封闭，密封

attrition *n.* 摩擦，磨损，消耗

M7-19
单词读音及例句

Reading comprehension

1. Can you summarize the advantages of fluidized bed membrane reactors?
2. Can you list the limitations of fluidized bed membrane reactors?

第八单元　化工仪表控制

Unit 8　Chemical Process Instrument and Control System

Lesson one: Measurement Instrumentation

In this lesson you will learn:

- Pressure measurement
- Flow measurement
- Level measurement
- Temperature measurement

1. Pressure measurement

The most widely used pressure unit in technical chemistry in continental Europe is the bar, while in China Pa is preferred. The conversion factors follow:

$$1\ \text{bar} = 10^5\ \text{Pa} = 0.987\ \text{atm}$$

Pressures can be specified as:

- Absolute pressure (P_a),
- Overpressure　$P_e = P_a - 1.013$ bar,
- Pressure difference DP.

As illustrated in Figure 8-1, three types of pressure-measuring devices can be distinguished on the basis of measurement principle:

- Pressure-measuring devices with a confining liquid (*e.g.*, U-tube manometer).
- Pressure-measuring devices with elastic pressure sensors.
- Pressure transducers.

continental　*adj.* 大陆的

absolute pressure 绝对压力

overpressure 超压（表压）

confine　*v.* 局限于，限制

manometer　*n.* 压力计

elastic　*adj.* 弹性的

transducer　*n.* 传感器

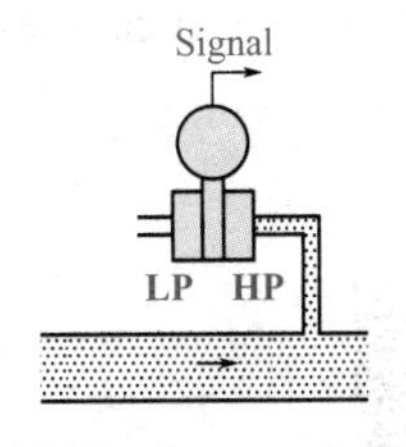

(a) DP cell > atmospheric

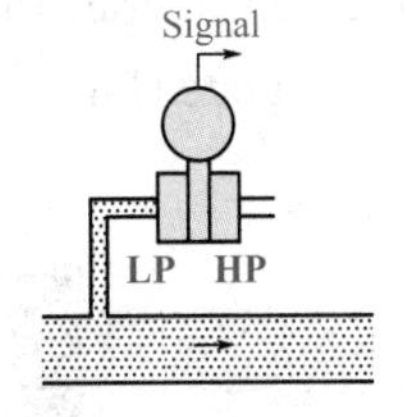

(b) DP cell for vacuum

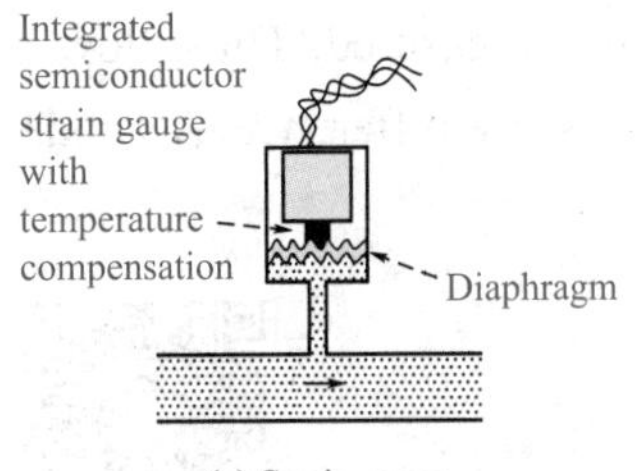

(c) Strain gauge

Figure 8-1

M8-1
单词读音及例句

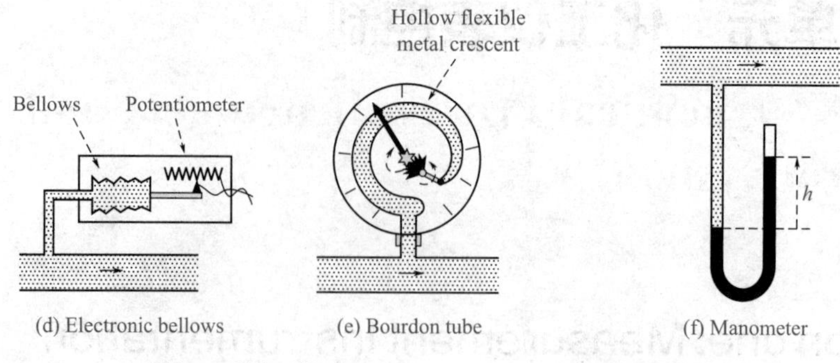

Figure 8-1 Various pressure measurement techniques

spring *n.* 弹簧
pulsating *adj.* 脉动的
piston pump 活塞泵
damping *n.* 衰减，阻尼
deform *v.* 使变形
inductive *adj.* 电感的
capacitive *adj.* 电容的
piezoresistive *adj.* 压阻的

While U-tube manometers for routine measurements are only rarely used for safety reasons, pressure-measuring devices with elastic pressure sensors (spring-tube, spring-plate, and capsule-spring manometers) are the most widely used because they are very robust and cheap.

In the case of pulsating pressures (*e.g.*, downstream of a piston pump) liquid-filled manometers must be used (damping). The use of a transducer, which affords an electrical measurement signal, is also possible. The measuring element of a pressure transducer is a metal membrane that is slightly deformed by changing pressure. This deformation can be transformed into an electrical signal by an inductive, capacitive, or piezoresistive transducer. Pressure transducers are much more expensive than manometers and are therefore preferentially used for transmitting measured data and measuring pressure differences. (From G. Herbert Vogel，Process Development 'From the Initial Idea to the Chemical Production Plant', 2005, 477)

2. Flow measurement

Flows are of fundamental importance for controlling a plant and for balancing and metering liquids. A large number of measurement principles are available here. Apart from the Coriolis-force meter, which directly measures the mass flux, all of them measure the volume flow.

(1) Throttle devices

flux *n.* 流量
throttle device 节流元件
orifice plate 孔板
jet *n.* 喷嘴

A mechanical resistance is introduced into the flow and the resulting pressure drop is measured. Thus, flow measurement is transformed into the measurement of pressure difference. A standard orifice plate (Figure 8-2) is generally used.

M8-2
单词读音及例句

M8-3
单词读音及例句

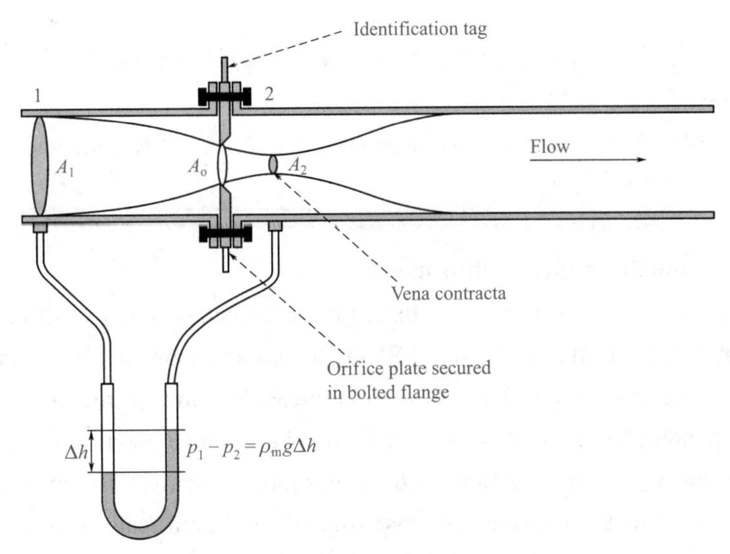

Figure 8-2 Orifice plate flow meter

Other types of throttle are the standard jet, and the standard Venturi jet, which have the advantage over the standard orifice plate of a lower permanent pressure drop. This measurement principle is applicable over a wide measuring range, requires little maintenance, is cheap, and is applicable for gases and liquids. The disadvantages are high permanent pressure drop (compression energy) and problems with difficult media (viscous liquids, media that cause fouling, multiphase flows).

viscous *adj.* 黏性的

fouling *n.* 污染，结垢

multiphase *adj.* 多相的

floating-body *n.* 浮子，转子

M8-4
单词读音及例句

(2) Floating-body flow meter

The use of these devices is restricted to the measurement of small to medium volume flows of low molecular weight liquids free of solid components (Figure 8-3). The position of the floating body can be measured inductively, so that an electrical signal is available for processing.

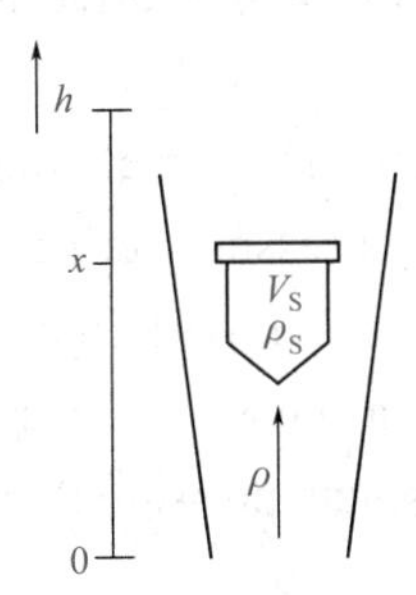

Figure 8-3 Floating-body flow meter

Notes:

The use of these devices is restricted to the measurement of small to medium volume flows of low molecular weight liquids free of solid components.

译文：这些设备的使用仅限于测量不含固体成分的低分子量液体的中小体积流量。

语法：句中 free of solid components 为后置定语，修饰 liquids, free of 意为"无……的"。

(3) Magnetic-induction flow meter

magnetic　*adj.* 有磁性的

induction　*n.* 感应

perpendicular to 垂直于

electrical potential 电势

M8-5
单词读音及例句

This measurement principle requires the presence of liquids with a certain minimum conductivity (at least 1 μS/cm; for comparison: distilled water has approx. 10 μS/cm). The volume flow to be measured flows through a magnetic field perpendicular to the direction of flow. The induced electrical potential is recorded by two electrodes and used as measured quantity. The measurement is independent of the pressure and viscosity of the liquid, causes no additional pressure drop, and can be used over a wide measurement range (Figure 8-4). (From G. Herbert Vogel, Process Development 'From the Initial Idea to the Chemical Production Plant', 2005, 477)

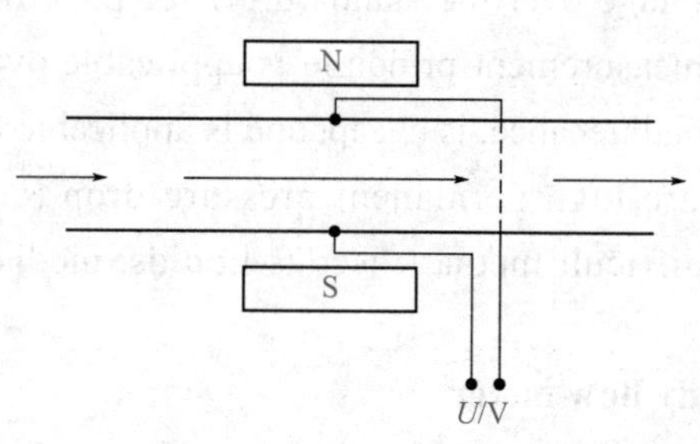

Figure 8-4　Magnetic-induction flow meter

3. Level measurement

Differential pressure (Figure 8-5) is often used to determine level. Three scenarios are shown below:

Measurements of liquid level for vessels open to atmosphere are obtained from $\Delta p=\rho gh$, with atmospheric pressure acting on both the liquid surface and the LP (Low Pressure) port of the DP (Differential Pressure) cell. For enclosed vessels, it is necessary to locate a second sensing point above the liquid surface, to eliminate the effect of any absolute pressure fluctuations.

Notes:

For enclosed vessels, it is necessary to locate a second sensing point above the liquid surface, to eliminate the effect of any absolute pressure fluctuations.

译文：对于封闭容器，必须在液面上方设置第二个传感点，以消除任何绝对压力波动的影响。

语法：be necessary to do sth，后接动词原型，意为"有必要做某事"。

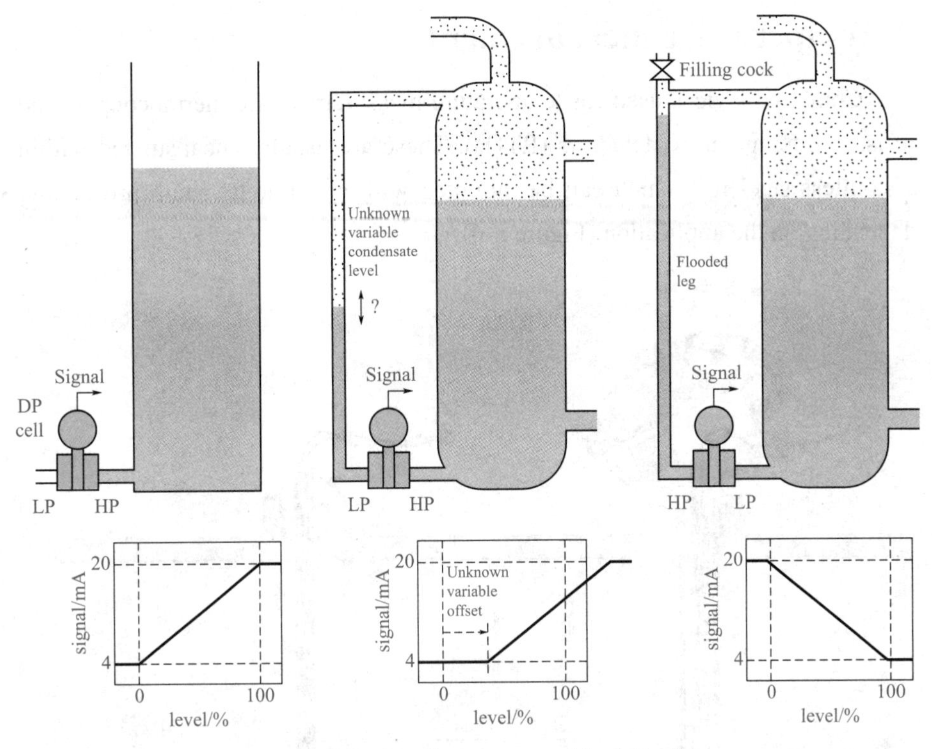

Figure 8-5 Level measurement by DP cell

This will involve an extra vertical section of impulse tubing, which introduces the problem of possible condensate collection, for example if the tubing temperature falls below that of the vessel. A serious situation can arise where the actual liquid level is much higher than that indicated. In practice, this problem is dealt with by reversing the DP cell port allocation, and ensuring that the contents of the high leg are known. Typically, the high leg is preloaded with the same liquid as in the vessel. A different, less volatile liquid might be used to avoid re-evaporation as process conditions change.

vessel *n.* 容器

fluctuation *n.* 波动，起伏

condensate *n.* 冷凝物

It is clear that levels obtained from $h=\Delta p/\rho g$ will be dependent on the average density of the liquid. Where this varies much, or froth forms or bubbles are present, procedures need to make allowance for the possible deviation, or an alternative level measurement technique must be sought. (From Michael Mulholland, Applied Process Control, 2016, 457)

reverse *v.* 颠倒，反过来

preload *v.* 预加载，预加料

re-evaporation *n.* 重蒸发

froth *n.* 泡沫

M8-6
单词读音及例句

M8-7
单词读音及例句

4. Temperature measurement

Common devices used for temperature measurement are thermocouples and resistance temperature detectors (RTDs). These are usually encapsulated within a stainless steel rod which can be mounted with or without sheath protection, depending on the application (Figure 8-6).

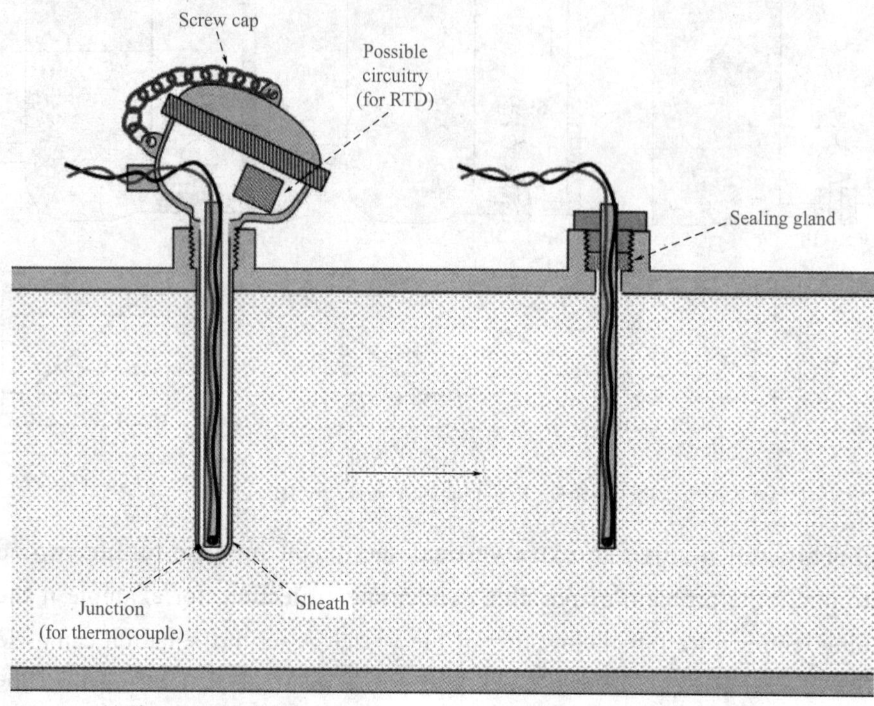

Figure 8-6　Sheathed and unsheathed temperature probes

thermocouple　*n.* 热电偶

encapsulate　*v.* 封装，包在荚膜内

mount　*v.* 安装

sheath　*n.* 鞘，护套

solder　*v.* 焊接

voltage　*n.* 电压

aging　*n.* 老化

M8-8
单词读音及例句

(1) Thermocouple Temperature Measurement

A thermocouple consists of two wires made of different materials (*e.g.*, Fe and CuNi) that are generally soldered together at one end (Figure 8-7). At the free wire ends (cold junction) a temperature dependent thermoelectric voltage can be measured.

The measuring probe is inserted in a protective sleeve and can be changed while the process is running. The measurement signal is transformed into a unit signal (4-20 mA) in a transducer. The overall error in the measurement is 1%. Common sources of error include variations in the thermoelectric voltage due to aging and drift of the cold junction. The preferred measuring range lies at higher temperatures (between 400 and +1000 ℃).

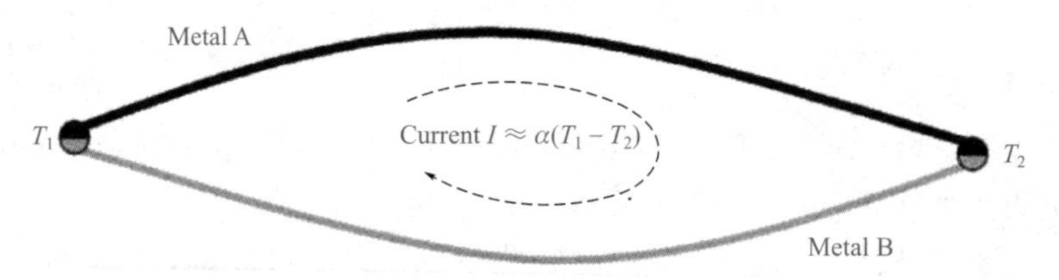

(a) Loop current due to junction emf(electromotive force) difference

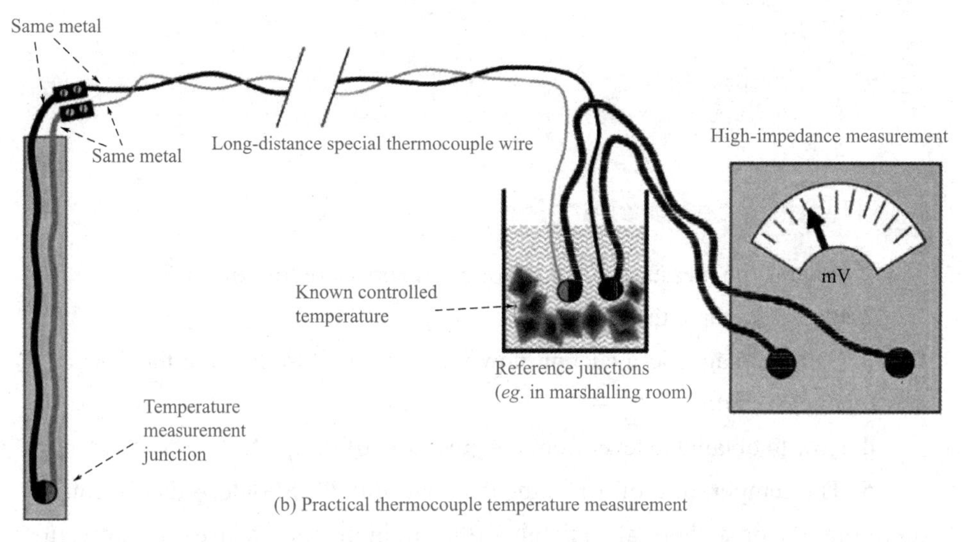

(b) Practical thermocouple temperature measurement

Figure 8-7　Principle of thermocouple temperature measurement

(2) Resistance temperature measurement

In the resistance thermometer the resistance of a metal wire coil (often Pt) is measured, and the temperature is determined from the known resistance/temperature function (so-called Pt 100, *i.e.*, 100 Ω at 0 ℃). The measuring probe is inserted in a protective sleeve and can be changed while the process is running. The measurement signal is transformed into a unit signal (4-20 mA) in a transducer. The overall error in the measurement is 0.5%. Depending on the required accuracy, two-, three-, or four-wire circuits are used. Common sources of error include corrosion of the clamping joint and poorly insulated protective sleeves. The preferred measuring range lies between −250 and +500 ℃.

(3) Bimetal thermometer

The thermometer, based on a bimetallic spiral in which strips of two dissimilar metals are fastened together, is the very common local measurement seen on plants. As the temperature rises, the two metals expand differently, imparting a turning moment in the spiral. One end of the spiral is fastened to the thermowell wall, whilst the free end rotates the needle on the dial.

resistance　*n.* 电阻
coil　*n.* 线圈，卷
clamping joint　钳位接点
bimetal　*adj.* 双金属的
spiral　*n.* 螺旋
turning moment　转矩
thermowell　*n.* 温度计套管
whilst　*conj.* 当……时候

M8-9
单词读音及例句

Notes:

One end of the spiral is fastened to the thermowell wall, whilst the free end rotates the needle on the dial.

译文：螺旋的一端固定在温度计套管壁上，而自由端则旋转表盘上的指针。

语法：句中 whilst 引导状语从句，和 while 含义基本相同，意为“当……的时候，和……同时；然而，而”。

Reading comprehension

1. What is the measuring element of elastic pressure transducers?

2. Please list three throttle devices.

3. Can the magnetic-induction flow meter be used to measure the flow of a non-conductive fluid?

4. How to obtain the level from the pressure difference ？

5. The temperature of a reactor is about 900 ℃. Should a thermocouple thermometer or a thermal resistance thermometer be used to measure the temperature?

Exercise

1. Fill in the blanks.

（1）1.013 bar =_______Pa =_______atm.

（2）_______ are the most widely used manometers.

2. Translate the following sentences into English or Chinese.

（1）The main problem is measuring the rather long time of passage of the ultrasound with sufficient accuracy.____________________

____

____

小提示：

- be restricted to 局限于
- be necessary to 对……来说有必要

（2）An standard orifice plate flow meter is restricted to the measurement of clean fluid.

____

____

（3）使用热电偶温度计时，有必要保证冷端温度的恒定。

____

____

Further Reading

Measurement Instrumentation

An initial division of measurement instruments is into the categories 'local' and 'remote'. A local device needs to be read at its point of installation, and has no means of signal transmission. Typical candidates are bimetal thermometer and Bourdon-type pressure gauges. Occasionally, manometers might be used to indicate differential pressures. Up until the 1980s, local measurements were prolific, and important, with operators patrolling the whole plant at hourly intervals, jotting down the readings on a clipboard. Increasingly now the operational picture is built up entirely from electronic information, whether it is updated on graphical mimic diagrams of the plant, or archived for future analysis. It seems that the additional investment in signal transduction, marshalling, conversion and capture is worthwhile in comparison with manual reading, transcription and data entry.

In the processing industries, there are few measurements requiring the high speeds of response often needed in electrical or mechanical systems. Nevertheless, it is important to bear in mind the impact of the response time constant of an instrument considered for each application. For example, a flue gas O_2 measurement might output a smoothed version of the actual composition. A brief O_2 deficiency might not be seen, but could be enough to start combustion when the uncombusted vapours meet O_2 elsewhere in the ducting.

prolific *adj.* 丰富的

patrol *v.* 巡逻

jot *v.* 略记，快速记载

mimic *adj.* 模拟的

archive *v.* 收集，归档

marshal *v.* 整理，集结

M8-10
单词读音及例句

deficiency *n.* 短缺

combustion *n.* 燃烧

M8-11
单词读音及例句

Reading comprehension

1. What is the difference between 'local' and 'remote' measurement instruments?
2. What is the importance of the response time constant of an instrument?

Lesson two: Controllers

In this lesson you will learn:

- Control technology
- Position controller
- Behavior of control circuits

1. Control technology

actuator *n.* 执行器

compensate *v.* 补偿

M8-12
单词读音及例句

Control here means that the measured value of a process variable. The so-called control variable x, is maintained at a given setpoint w by adjusting actuators such as valves and pumps by means of a control element until the deviation between x and w is compensated. The control command (controller output y) is thus always dependent on the setpoint (feedback). The principle of a simple control circuit is shown as Figure 8-8.

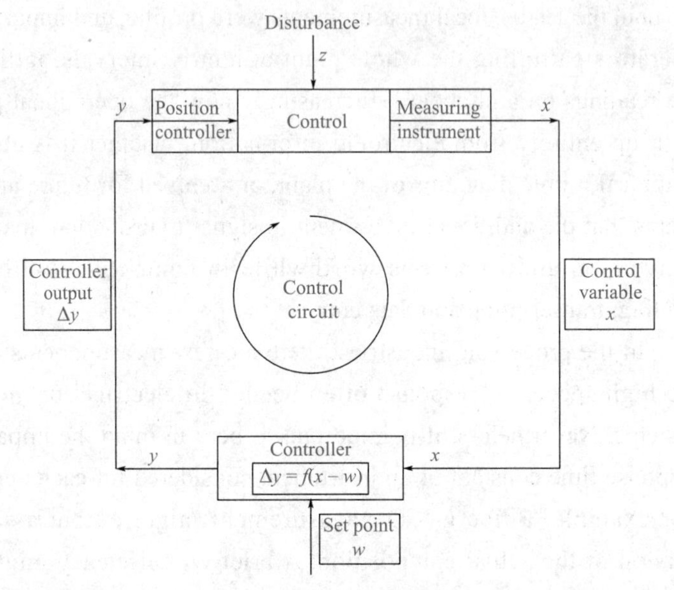

Figure 8-8　Basic elements of a closed-loop control circuit

These transform the output signal of a measuring instrument into a standardized signal (*e.g.*, 4-20 or 0-20 mA, 0.2-1 bar).

2. Position controller

circuit *n.* 回路

temporal *adj.* 暂时的

M8-13
单词读音及例句

Position controller mechanically measures the true valve position *in situ* and compares the value with the controller output y provided by the controller. In cases of control deviation, the control element is adjusted until the required position y is reached.

(1) Time behavior of control circuits (important: controller is in manual mode)

Knowledge of the dynamic behavior of the process to be controlled (control circuit), that is the temporal behavior of the control variable $x(t)$ resulting from a sudden change in the controller output Δy is important for the design of the entire control circuit. This behavior can be analyzed by suddenly increasing the controller output y by the amount Δy (*e.g.*, by opening the control valve from 50% to 60%) and recording the value x. Figure 8-9 (a) shows the increase in controller output y by the amount Δy, while Figure 8-9 (b) indicates the response $x(t)$.

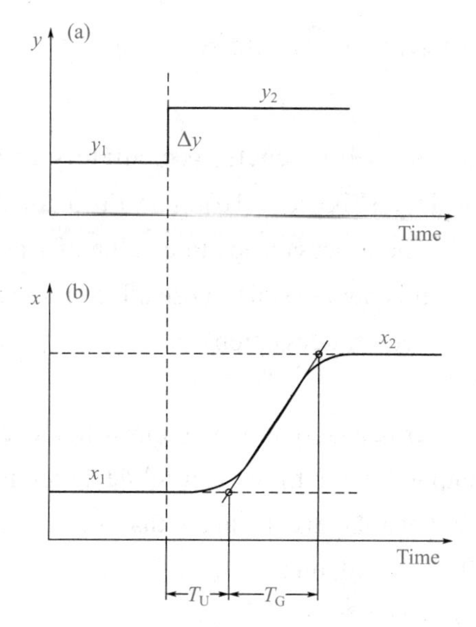

Figure 8-9　Dynamic behavior of a control circuit:
(a) Increase in controller output y by the amount Δy; (b) Response $x(t)$

A tangent to the turning points can be used to determine the two characteristic quantities delay time T_U and compensation time T_G. When both values are zero, the control circuit is without delay or of zeroth order, and when only T_G is zero the circuit is first-order.

tangent　*n.* 切线
zeroth order　零级

(2) Time behavior of controller types (important: without control circuit)

The controller type or the parameters to be adjusted on the controller must be chosen or optimized in accordance with the expected dynamic behavior of the control circuit. An important characteristic of a controller is the temporal variation of the controller output in response to a sudden control deviation. According to this behavior, three types of controller are distinguished: P, I, and D controllers. (From G. Herbert Vogel, Process Development 'From the Initial Idea to the Chemical Production Plant', 2005, 477)

M8-14
单词读音及例句

Notes:

The controller type or the parameters to be adjusted on the controller must be chosen or optimized in accordance with the expected dynamic behavior of the control circuit.

译文：控制器类型或要在控制器上调整的参数必须根据控制电路的预期动态行为进行选择或优化。

语法：句中 to be adjusted 为不定式，作后置定语，修饰 the parameters；词组 in accordance with 意为“按照……，依据……”。

3. Behavior of control circuits

(1) P controller

The proportional controller reacts immediately to an $x-w$ jump with a proportional change in y (Figure 8-10). The ratio $\Delta w/\Delta y$ is known as the proportionality range X_p. The process leads to a value of x that generally deviates from the setpoint, that is, it is not possible to use a P controller to exactly control x to the setpoint (permanent control deviation).

proportional *adj.* 比例的
permanent control deviation 余差
integral *adj.* 积分的
sluggish *adj.* 迟缓的
oscillation *n.* 振荡
inherent *adj.* 固有的，内在的
combine *v.* 组合

M8-15
单词读音及例句

(2) I controller

The integral controller reacts to an $x-w$ jump with a continuously increasing change in controller output $\Delta y(t)$ until the control deviation becomes zero (Figure 8-10). Pure I controllers have the disadvantage that they are either too sluggish or are prone to uncontrollable oscillations.

(3) D controller

The differential controller reacts to a control deviation with a y pulse, that is the controller output y rises sharply and then returns slowly to the initial value, so that no inherent control function is present (Figure 8-10).

In practice the basic types are often combined to achieve optimum control behavior. The most important combinations that are used in chemical plants are PI controller and PID controller.

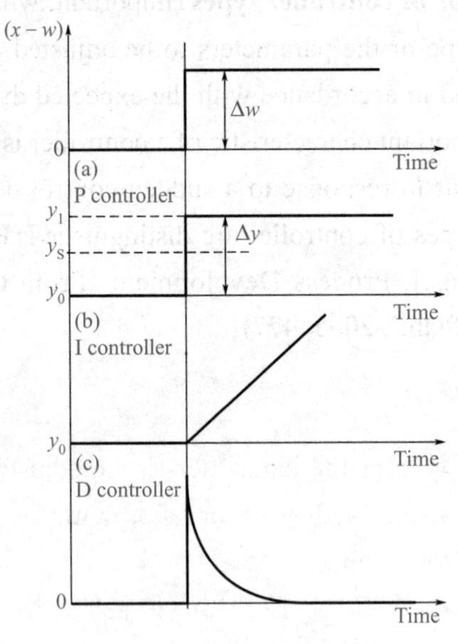

Figure 8-10 Temporal variation of the controller output y in response to a sudden control deviation $x-w$ for the three basic types of controller

(4) PI controller

Overall this controller has a favorable control behavior (no permanent control deviation). It acts immediately, but exhibits relatively slow adjustment to the setpoint. The time behavior of the PI controller is characterized by the reset time T_N. It is especially suitable for pressure and flow control circuits.

(5) PID controller

The PID controller combines the advantages of the three basic types: very fast action and fast adjustment to the setpoint without a permanent control deviation. It is especially suitable for temperature control. An ideal PID controller can be described mathematically by Equation:

$$\Delta y = \left\{ K_P \cdot \Delta x + \frac{1}{T_N} \cdot \int_{t=0}^{t} \Delta x \cdot dt + T_V \cdot \frac{d\Delta x}{dt} \right\}$$

The individual contributions of the three basic types to a PID controller and hence the control behavior are characterized by three characteristic values:

- K_P: proportional range, changing P behavior
  (small K_P value = large P contribution).
- T_N: reset time, changing I behavior
  (small T_N value = large I contribution).
- T_V: derivative action time, changing D behavior
  (large T_V value = large D contribution).

Note that the three quantities are not independent of one another. For optimal adaption of the controller to the control circuit, the behavior of both must be known. Random adjustment of all three parameters is generally not successful. A series of methods are available for determining favorable control parameters, whereby on the basis of targeted adjustments and the resulting reaction of the control circuit, the correct setting of the controller can be found.

If the measured quantity of the control circuit undergoes periodic oscillations, then control is unstable. The reason for this usually lies in too strong an action of the controller. Here it often helps to enlarge the proportional range (smaller P contribution) or to increase T_N. With a PID controller one must ensure that T_N is always at least four to five times T_V. (From G. Herbert Vogel, Process Development 'From the Initial Idea to the Chemical Production Plant', 2005, 477)

proportional range
比例度

derivative *n.*
导数，微分

whereby *adv.*
凭借，因此

periodic *adj.*
周期的

M8-16
单词读音及例句

Reading comprehension

1. Please describe the principle of a simple control circuit.
2. How many controller types can be adjusted on the controller?

Exercise

1. True or False.

(1) The controller output y would increase immediately respond to a change in $x(t)$.

（2）A large proportional range means a large P contribution.

（3）Each contributions of the three basic types to a PID controller is individual and hence the control behavior is characterized by three characteristic quantities. So the three quantities are independent of one another.

2. Translate the following sentences into Chinese.

（1）Control is a process in a system in which input quantities influence output quantities on the basis of a logical operation.______________________

____

（2）It takes into account the signal states of the inputs and, depending on the computational operation, assigns the outputs.______________________

____

Further Reading

Controllers

panel-mounted 面板安装

analogue *adj.* 模拟的

terminology *n.* 术语

algorithm *n.* 算法

M8-17 单词读音及例句

Until the 1980s, panel-mounted analogue controllers were common. Sometimes these were integrated within a circular or strip chart recorder. This is the SISO (single-input single-output) controller which could service a single control loop. With the conversion to computer-based systems, the same concepts and terminology have been retained. The same functions are carried out by programs which run on PLCs (programmable logic controllers) or distributed computer cards, and these can continue to function independently if part of the system is damaged. When the operator views a controller on the control room displaying of a DCS (distributed control system) or SCADA (supervisory control and data acquisition) system, it is represented in analogous format. On the conversion of plants to the much cheaper computer-based systems, the control loops were replaced on a one-for-one basis. The many tasks requiring this type of control are referred to as the 'base layer' control system, and it will be seen later that the modern advanced control algorithms facilitated by computers tend to communicate with the plant through this layer.

The loop controller has a switch to select either AUTO or MANUAL operation (Figure 8-11). In the MANUAL mode, it is possible to adjust the control action output directly, using the 'manual loading station' at the bottom of the controller. In AUTO mode, the device makes use of its internal computation to set a suitable control action output which will bring the feedback signal to the setpoint value.

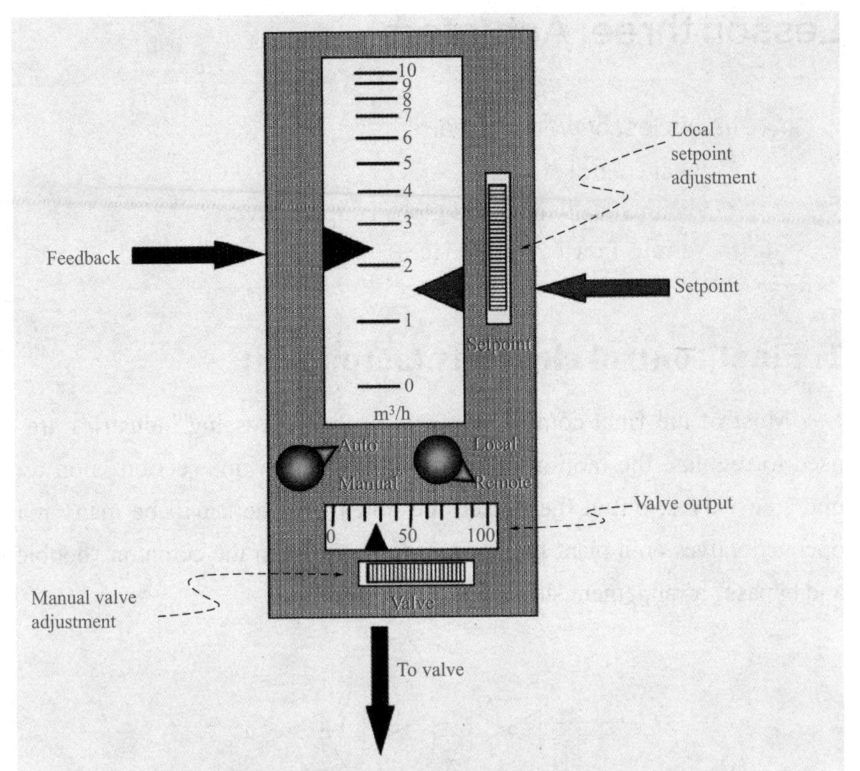

Figure 8-11　Typical panel-mounted analogue controller

The other switch on the controller allows selection of LOCAL or REMOTE. In LOCAL mode, the setpoint can be adjusted directly by the operator. In REMOTE mode, this is no longer possible, because the setpoint is manipulated from elsewhere. This could be in a cascade format from a similar SISO controller, or possibly from an 'advanced' control algorithm based on several PVs (process variables) and manipulating several MVs (manipulated variables).

cascade　*n.* 串联，串级

M8-18
单词读音及例句

The original analogue controllers were used in one of the three control modes: proportional(P), proportional-integral (PI) and proportional-integral-derivative (PID). Rather more features are possible today in the computer equivalent controller. For the analogue devices, one accessed the back of the cabinet to adjust the proportional gain, integral time constant or derivative time constant. These can of course be set remotely in the computer devices. (From Michael Mulholland, Applied Process Control, 2016, 457)

Reading comprehension

1. What is the difference between AUTO or MANUAL operation?
2. How to adjust the control modes on the original analogue controllers?

Lesson three: Actuators

In this lesson you will learn:

- Final control elements
- Pneumatically operated globe control valve
- Various flow control devices

1. Final control elements (actuators)

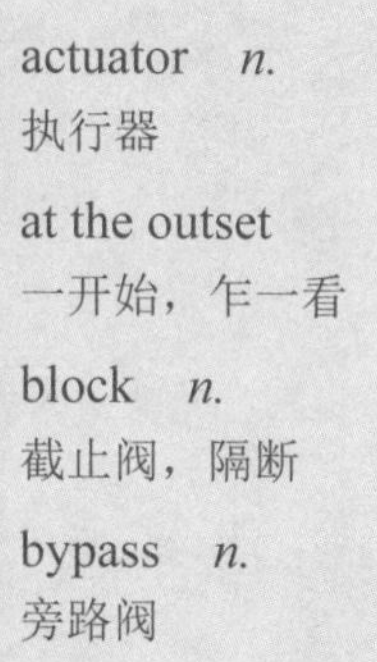

Most of the final control elements in the processing industries are valves used to regulate the motion of fluids. The focus in this section is on remotely operated valves, but at the outset one notes that there will be many manually operated valves on a plant for less frequent use, as in the common "double-block and bypass" arrangement shown in Figure 8-12.

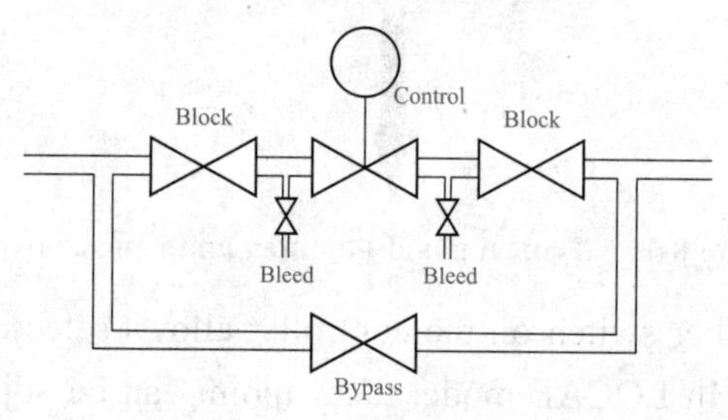

Figure 8-12　The common "double-block and bypass" arrangement

debottlenecking
消除瓶颈
bleed cock
放空阀
rupture　*v.* 使破裂

Control valves sometimes need maintenance, so that they are often located in accessible spots near floor level, to allow disconnection and removal. A means must be provided to prevent residual process fluids from escaping through the vacancy that is the purpose of manual block valves. Quite often, with communication to an operator at the location, the process can continue to function temporarily with occasional manual adjustments of the bypass valve. A quite undesirable practice is sometimes seen where debottlenecking of a process has led to bypass valves being left partially open in normal operation, just to increase throughput. In addition to the three manual valves involved, two manual bleed cocks will sometimes be provided in liquid systems not only to empty the closed-off sections in a controlled way, but also to avoid trapping of liquid which on thermal expansion might rupture the pipework.

Notes:

In addition to the three manual valves involved, two manual bleed cocks will sometimes be provided in liquid systems not only to empty the closed-off sections in a controlled way, but also to avoid trapping of liquid which on thermal expansion might rupture the pipework.

译文：除了所涉及的三个手动阀外，有时在液体系统中会提供两个手动放空阀，不仅能以可控方式清理堵塞部分，而且还可以避免因热膨胀而导致管道破裂的液体滞留。

语法：句中包含关联词组 not only... but also，意为“不仅……，而且……”；句中关系代词 which 引导定语从句，修饰 trapping of liquid。

2. Pneumatically operated globe control valve

The globe control valve is the most common device used for precise flow regulation, and the simple and safe source of motive power for this valve is usually the instrument air system.

Unlike domestic taps, note that the flow normally approaches from the stem side of the valve plug. An adjustable stuffing box with a gland prevents leakage to the outside. Most control valves can be switched between air-to-open and air-to-close service by simple mechanical adjustments (Figure 8-13). Bearing in mind that the most likely disaster would be loss of instrument air pressure, or breakage of the pneumatic signal line, the air-to-open (AO) is taken as fail-closed (FC) and the air-to-close (AC) is taken as fail-open (FO). In control scheme design, it is important to make this specification to ensure as orderly a shutdown as possible in the event of disaster, for example furnace fuel supplies will typically be fail closed. For motor operated valves, or electrical solenoid valves, the fail situation would correspond to loss of the electrical power or signal.

pneumatically *adv.* 充气地

precise *adj.* 精确的

domestic *adj.* 家用的

stuffing box 填料函

gland *n.* 密封压盖

furnace fuel 炉用燃料

motor-operated 电动操纵的

solenoid valve 电磁阀

M8-21
单词读音及例句

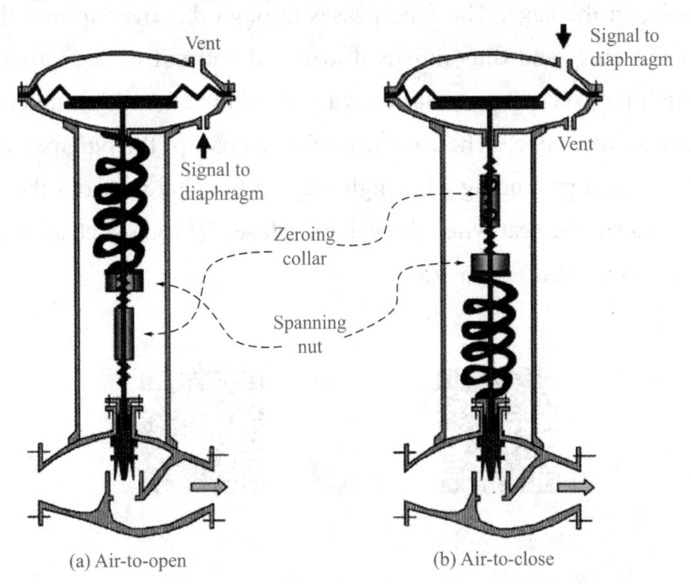

Figure 8-13 Valve air-to-open and air-to-close arrangements with adjustments

Notes:

Bearing in mind that the most likely disaster would be loss of instrument air pressure, or breakage of the pneumatic signal line, the air-to-open (AO) is taken as fail-closed (FC) and the air-to-close (AC) is taken as fail-open (FO).

译文：考虑到最可能发生的灾难是仪表风损失或气动信号线损坏，气开型（AO）被视为出故障时自动关闭（FC），气关型（AC）被视为出故障时自动打开（FO）。

语法：Bearing in mind that... signal line 为现在分词短语做状语；that 为连词，引导宾语从句。

collar *n.* 圈，箍，衣领

stem *n.* 阀杆

nut *n.* 螺母

A collar allows variation of the stem length, whilst one or two nuts may vary the position and tension on the return spring. It is important to check these adjustments, so that the actual motion of the stem between 0% open and 100% open corresponds as well as possible to the instrument signal range, say 3 and 15 psig (or 15 and 3 psig for an air-to-close valve, 1psig=6894.76 Pa). A point of caution is that valve signals are occasionally reported as '% closed', so it is wise to qualify the units of measurement as either '% open' or '% closed' when valve position is being discussed.

circumference *n.* 圆周

overlap *n.* 重叠，交错

hollow *adj.* 中空的

perforated *adj.* 穿孔的，有排孔的

protruding *adj.* 凸出的

seal *n.* 密封

The arrangement inside a globe control valve with cage and plug is illustrated in Figure 8-14. The arriving fluid passes through the holes in the cage, around its full circumference. The plug passes up and down within the cage, with the various holes cut into the plug's cylindrical surface overlapping in a predetermined way with the holes in the cage. The fluid passes through this overlap into the hollow interior of the plug, and thus moves downward and out of the valve. It is the way in which the overlap area varies with stem movement that determines the characteristic of the valve. Whereas the perforated part of the plug passes through a hole in the seat, a protruding edge higher up on the plug provides the means for a tight seal against the seat when the valve is closed. (From Michael Mulholland, Applied Process Control, 2016, 457)

M8-22
单词读音及例句

M8-23
单词读音及例句

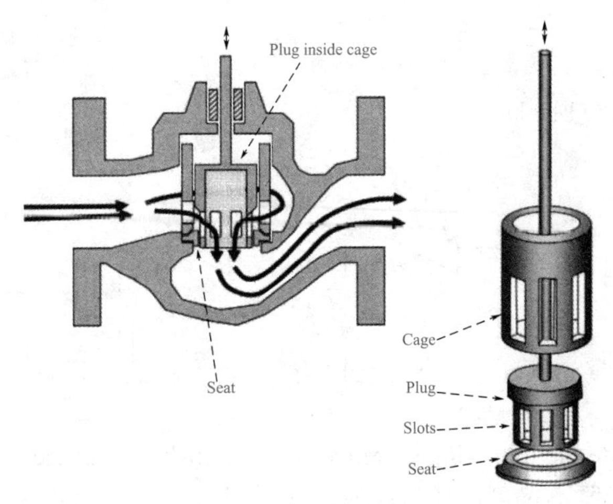

Figure 8-14　Globe valve showing fluid flow through cage and plug

3. Various flow control devices

A great variety of flow control devices exists, suited to different applications. Some are discrete in the sense that they only use two positions, some are designed for accurate regulation and some are designed to handle fluids containing solid particles. The diaphragm valve can be used for particulate-bearing flows, because it is able to flush itself clear. Gate or ball valves can be arranged to leave the entire pipe flow area clear, and are thus suited to special cleaning requirements. Louvres and baffles would be suited to low-pressure systems such as furnace ducting.

diaphragm *n.* 隔膜阀

louvre *n.* 百叶窗

baffle *n.* 挡板

ducting *n.* 管道，导管

M8-24
单词读音及例句

Notes:

Some are discrete in the sense that they only use two positions.

译文：有些是离散的，从某种意义上说它们只使用两个位置。

语法：句中连词 that 引导同位语从句，in the sense that 意为“从某种意义上说”。

Reading comprehension

1. What is the purpose of the “double-block and bypass” arrangement?

2. How to determine which actuator should be used in the control system, air-to-open or air-to-close?

Exercise

1. Fill in the blanks.

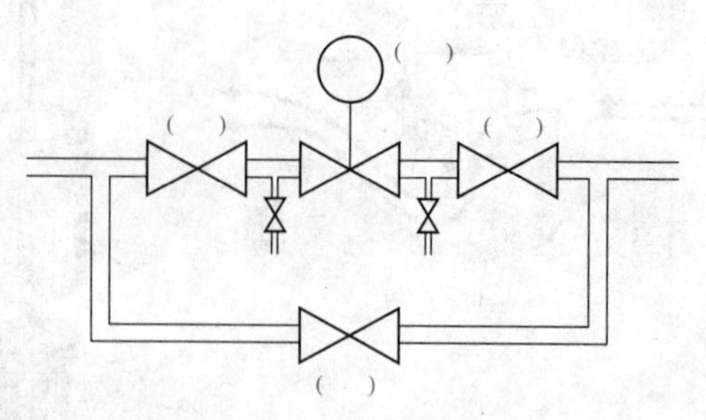

2. Translate the following sentences into English or Chinese.

（1）执行器不仅可以通过控制系统自动操作，也可以手动操作。________________

____

（2）In the sense that it only needs a slight modification to the air-to-open actuator to become an air-to-close actuator.________________

____

小提示：

• not only... but also...
不仅……而且……

Further Reading

Actuator Systems

Together with sensor systems, actuator systems link material, energy, and information streams in processing and manufacturing systems. The control path with sensor system, controller, and actuator system forms a closed control loop. Changes, whether in perturbing values, setpoints, controlled variables, or other quantities, manifest themselves in the system and also in the signals transmitted to and by these devices. With the aid of actuator systems, the material or energy stream is set in the control loop as a function of the system deviation. The action also depends on the characteristics of the control path, the sensor system, and the controller with its control parameters influencing its dynamics. With regard to the dynamics, the control element can be included in the control pathway and the servo drive (actuator drive) can be included in the control device. In the open loop case, the intervention in the process is effected through a control command, which takes the form of a control action or a manual action. The control pathway need not carry feedback.

In actuator systems, the control command (the output of the controller) must be transformed to a control signal. Servo drives in the form of power amplifiers are needed for this purpose. These may be pneumatic, electric,

perturbing value
干扰值

manifest　*v.* 表明

servo drive
伺服驱动

intervention　*n.*
介入，调停

M8-25
单词读音及例句

or hydraulic, but hydraulic drives are rarely used in process engineering. Because of their well known advantages in manufacturing and robotics, however, hydraulic power amplifiers are preferred in these fields. They are robust drives whose high power density results in good control of complex motions, accurate positioning under load, and wide speed ranges. Problems, however, such as poor damping, marked nonlinearity, and load-dependent eigenfrequencies, should be mentioned.

Continuous and switching control elements are distinguished, just as control elements with memory must be distinguished from those with a well-defined rest position. Single action modes or combinations may be required, depending on process or safety requirements. Continuous control elements affect a steady or gradual change in mass or energy flow within the design range, while switching elements act in a step-by-step manner. In process control engineering, there is often only a minor difference between the two types. Control elements of continuous or switching type can be designed by modification of globe and plug valves, cocks, shutoff valves, and the like. The required characteristic determines whether the required accuracy of positioning or control can be achieved under given pressure conditions. The characteristic associated with the system (inherent characteristic) can be matched to the operating characteristic. With regard to accuracy of positioning, a corresponding resolution of the control signal must be ensured. Control elements such as slide valves or solenoid-actuated valves have a design that makes them unsuitable for continuous control and positioning.(From M. Polke, Process Control Engineering, 1994, 478)

hydraulic *adj.* 液压的

amplifier *n.* 放大器

eigenfrequency *n.* 本征频率

minor *adj.* 小的

M8-26
单词读音及例句

Reading comprehension

1. Hydraulic drives are rarely used in process engineering, so what are their advantages?

2. Describe the difference between continuous and switching control elements.

Lesson four: Process Control

In this lesson you will learn:

- Feedback control
- Feedforward control
- Cascade control

1. Feedback control

Feedback control is a very important aspect of process control. Its role is best described in terms of an example as shown in Figure 8-15.

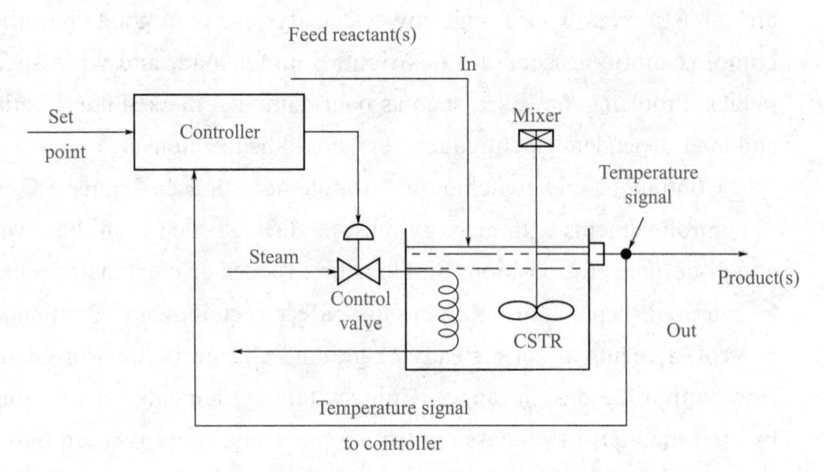

Figure 8-15　Feedback control system of a CSTR

feedback *n.* 反馈
relay *v.* 转送
comparator *n.* 比较器
magnitude *n.* 大小，量级
criterion *n.* 准则

M8-27
单词读音及例句

Assume that one desires to maintain the temperature of a polymer reactor at 70 ℃. Temperature is thus the controlled variable, and the desired temperature level 70 ℃, is called the aforementioned set-point. In feedback control, the temperature is measured using a sensor (such as a thermocouple device). This information is then continuously relayed to a controller, and a device known as a comparator, compares the set-point with the measured signal (or variable). The difference between the set-point and the measured variable is the previously defined error. Based on the magnitude of the error, the controller element in the feedback loop takes corrective action by adjusting the value of a process parameter, known as the manipulated variable (Figure 8-16).

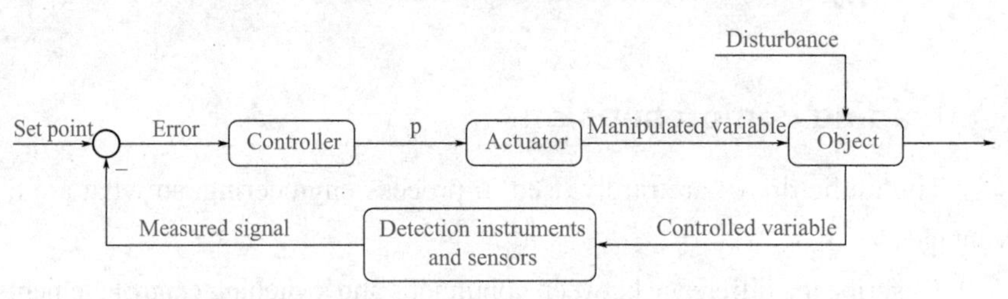

Figure 8-16　Block diagram of a simple feedback control system

The controller logic (the manner in which it handles the error) is an important process control criterion. Generally, feedback controllers are proportional (P, sends signals to the final control element proportional to the error), proportional-integral (PI, sends a signal to the final control element that is both proportional to the magnitude of the error at any instant and the sum of the error), and proportional-integral-derivative (PID, sends a signal that is also based on the slope of the error).

In the above example, the manipulated variable may be cooling water flow through the reactor jacket. This adjustment or manipulation of the flow rate is achieved by a final-control-element (actuator). In most chemical processes, the actuator is usually a pneumatic control valve. However, depending upon the process parameter being controlled, the final-control-element could very well be a motor whose speed is regulated. Thus, the signal from the controller is sent to actuator which manipulates the manipulated variable in the process.

manipulated variable 操作变量

controlled variable 被控变量

load disturbances 负荷扰动

M8-28
单词读音及例句

Notes:

However, depending upon the process parameter being controlled, the final-control-element could very well be a motor whose speed is regulated.

译文：然而，取决于要调节的工艺参数，最终控制元件很可能是一种转速可调的电机。

语法：depending... controlled 为现在分词短语作句子的状语；句中 whose 引导定语从句，修饰 motor。

In addition to the controlled variable, there may be other variables that disturb or affect the process. In the reactor example above, a change in the inlet temperature of the feed or inlet flow rate is considered to be load disturbances. A servo problem is one in which the response of the system to a change in set-point is recorded, whereas a load or regulator problem is one in which the response of a system to a disturbance or load variable is measured.

Before selecting a controller, it is very important to determine its action. Consider another example—the heat exchanger as shown in Figure 8-17.

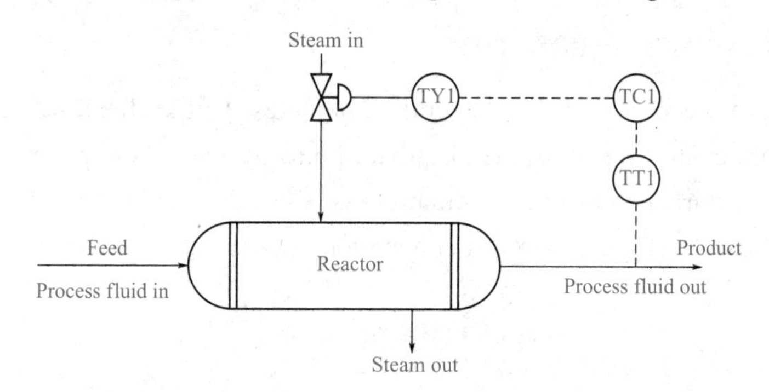

Figure 8-17　Feedback control system of heat exchanger

Steam is used to heat the process fluid. If the inlet temperature of the process fluid increases, an increase in the outlet temperature will result. Since the outlet temperature moves above the set-point (or desired temperature), the controller must close the steam valve. This is achieved by the controller sending a lower output (pneumatic or current) signal to the control valve, *i.e.*, an increase in the input signal from the controller to the valve. The action of the controller is considered to be reverse. If the input signal to the controller and the output signal from it act in the same direction, the controller is direct acting.

> **Notes:**
>
> If the input signal to the controller and the output signal from it act in the same direction, the controller is direct acting.
>
> 译文：如果控制器的输入信号和控制器的输出信号作用是同一个方向，则控制器是正作用的。
>
> 语法：句首连词 if 引导条件状语从句；to the controller 为后置定语，修饰 signal。

microprocessor *n.* 微处理器
scale factor 比例因子
twofold *adj.* 双重的，两方面
sole *adj.* 唯一

M8-29
单词读音及例句

It is important to consider the process requirements for control to determine the action of the controller and the action of the final control element. The controller action is usually set by a switch on electronic and pneumatic controllers. On microprocessor-based controllers, the setting can be made by changing the sign of the scale factor in the software (which then changes the sign of the proportional gain of the controller).

The functions of a feedback controller in a process control loop are twofold: (1) to compare the process signal from the transmitter (the controlled or measured variable) with the set-point, and (2) to send a signal to the final control element with the sole purpose of maintaining the controlled variable at its set point.

As noted above, the most common feedback controllers are proportional controllers (P control), proportional-integral controllers (PI control), and proportional-integral-derivative (PID) controllers.

2. Feedforward control

feedforward control 前馈控制
preemptive *adj.* 先发制人的
entail *v.* 需要，承担

M8-30
单词读音及例句

Feedforward control has several advantages. Unlike feedback control, a feedforward control measures the disturbance directly, and takes preemptive action before the disturbance can affect the process. Consider the heated tank example discussed earlier (Figure 8-18). A conventional feedback controller would entail

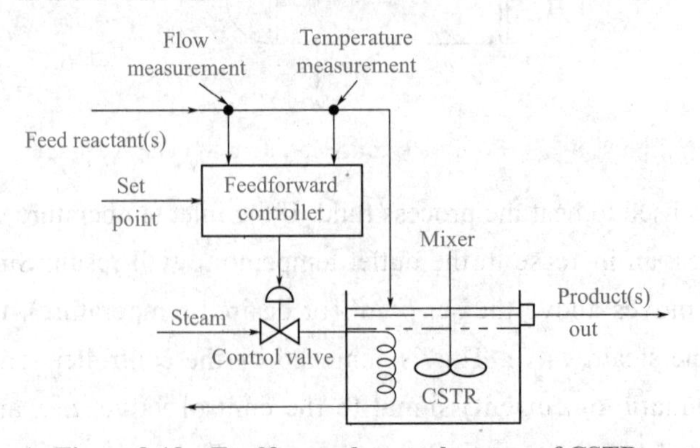

Figure 8-18 Feedforward control system of CSTR

measuring the temperature in the tank, the controlled variable, and maintaining it at the desired set-point by regulating the heat input, the manipulated variable.

In feedforward control, the load disturbances would first be identified, *i.e.*, the inlet flow rate, and the temperature of the inlet fluid. Any change in the inlet temperature, for example, would be monitored in a feedforward control system, and corrective action taken by adjusting the heat input (again the manipulated variable), before the process is affected. Thus, unlike feedback control, the error is not allowed to propagate through the system. It is clear that a feedforward controller is not a PID controller, but a special computing or digital machine.

propagate *v.* 传播，传送

cascade control 串级控制

M8-31 单词读音及例句

Notes:

It is clear that a feedforward controller is not a PID controller, but a special computing or digital machine.

译文：很明显，前馈控制器不是 PID 控制器，而是一种特殊的计算器或数字机器。

语法：句首的 it 为先行主语，代表后移的主语从句 that a feedforward controller…。

Good feedforward control relies to a large extent on good knowledge of the process, which is often the biggest drawback. Finally, the stability of a feedforward-feedback system is determined by the roots of the characteristic equation of the feedback loop (feedforward control does not affect the stability of the system).

3. Cascade control

The simple feedback control loop considered earlier is an example of a Single Input Single Output (SISO) system. In some instances, it is possible to have more than one measurement but one manipulated variable, or one measurement and more than one manipulated variable. Cascade control is an example where there is one manipulated variable but more than one measurement.

Consider a slight modification of the continuous stirred tank reactor (CSTR). The control objective is to maintain the reactor temperature at a set value by regulating the cooling water flow to the exchanger. The load disturbances to the reactor include changes in the feed inlet temperature or in the cooling water temperature. In simple feedback control, any change in the cooling water temperature will affect the reactor temperature, and the disturbance will propagate through the system before it is corrected. In other words, the control loop will respond faster to changes in the feed inlet temperature compared to changes in the cooling water temperature.

exchanger *n.* 交换器

secondary control loop 副回路

M8-32 单词读音及例句

Notes:

In other words, the control loop will respond faster to changes in the feed inlet temperature compared to changes in the cooling water temperature.

译文：换句话说，与冷却水温度的变化相比，控制回路对进料入口温度的变化响应更快。

语法：句中 compared to… temperature 为过去分词短语作句子的状语。

Now consider the following example in Figure 8-19. In this case, any change in the cooling water temperature is corrected by a new controller added to the process control loop before its effect propagates through the reaction system. These are two different measurements—reaction temperature and cooling water temperature, but only one manipulated variable (cooling water flow rate). The loop that measures the reaction temperature is known as the primary control loop, and the loop that measures the cooling water temperature, the secondary control loop. The secondary control loop uses the output of the primary controller as its set-point, whereas, the set-point to the primary controller is supplied by the operator. Cascade control has wide applications in chemical processes. Usually, flow rate control loops are cascaded with other control loops.

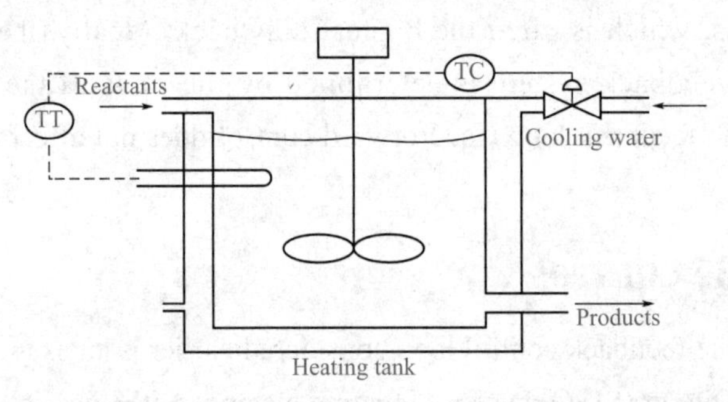

Figure 8-19 Cascade temperature control system of heating tank

Reading comprehension

1. What is the role of the comparator in feedback control?
2. What is the difference between feedback control and feedforward control?
3. How many control loops are there in cascade control?

Exercise

1. True or False.

（1）The most common feedback controllers are P controllers, PI controllers, and PID controllers.

（2）Feedforward control has advanced control capability.

（3）There are two manipulated variables in cascade control.

2. Translate the following sentences into English or Chinese.

（1）Feedforward control is likely to be suitable for a process where the disturbance changes frequently.____

____

（2）虽然串级控制结构更为复杂，但是它很有可能具有更好的控制效果。

____

____

Further Reading

Importance of Control in Chemical Processing

We will focus on modelling, estimation, control and optimisation in the processing industries. There are unique challenges here to do with the inaccuracy of models and undefined disturbances. In addition, the widespread use of computers to handle process instrumentation in recent decades has spurred the concept of ‘advanced process control’ (APC), which has become a specialised process engineering domain. The objective is to take advantage of the plant-wide view of outputs, and access to inputs, of these computers, in order to enhance regulation and optimisation. In this way, industries have been able to work safely with narrower specifications and less loss. With the increasing globalisation of markets, industries which do not seek such efficiency improvements will soon find themselves uncompetitive and out of business.

In the processing industries, the automatic control aspects are viewed to constitute a pyramid of three main layers in which each layer achieves its objectives by supervising the layer below. Generally, this means that the control loop setpoints (SPs) are passed downwards.

Usually the base layer becomes the responsibility of instrumentation technicians, but more advanced inputs are required from control engineers in the upper layers. Of course, the overall control scheme, including the base layer, must be specified by engineers in the design phase. At that stage, additional specifications may be made, such as increased vessel hold-ups to facilitate ‘advanced level control’. Indeed, there is a growing trend to integrate the equipment design and control design at an early stage. Increasing integration of processes through

optimisation *n.* 最优化

inaccuracy *n.* 不准确

spur *v.* 鞭策

constitute *v.* 构成

supervise *v.* 监督，照看

M8-33 单词读音及例句

render　*v.* 提出
algorithm　*n.* 算法
oversight　*n.* 监督
throughput　*n.* 吞吐量，生产量
downtime　*n.* 故障停机时间
transparent　*adj.* 易懂的，透明的

M8-34
单词读音及例句

'pinch' analysis often renders the internal regulation highly interactive, requiring special control approaches. Another lesson that has been learned is that the advanced control algorithms cannot simply be installed and left to operate without ongoing knowledgeable oversight.

The advanced algorithms focus on criteria such as throughput, product specifications and economics, not necessarily smooth process operation, and thus they can be unpopular with operating personnel. All too often such unwelcome behaviour can cause operators to switch off these algorithms. Thus, education is important, as well as investigation of downtime incidences and constant reviewing of performance. On one level, one aims to make a control scheme as simple and transparent as possible, to facilitate understanding. However, some algorithms are unavoidably complex, referring to a number of measurements as the basis of their output decisions. A specially trained control engineer is required to diagnose poor performance that might arise from a poorly calibrated measurement. Industries which recognise the need for ongoing care, and provide the necessary resources, have successfully increased the fractional online time of their optimisers and advanced controllers.

(From Michael Mulholland, Applied Process Control, 2016, 457)

Reading comprehension

1. What is the purpose of advanced process control?
2. Why are the advanced algorithms unpopular with operating personnel?

第九单元　工艺流程图

Unit 9　Chemical Process Diagram

Lesson one: Basic Flow Diagram

In this lesson you will learn:

- Flow diagrams
- Basic flow diagram

1. Flow diagrams

Flow diagrams describe in a schematic drawing format the flow of fluids and gases through a unit or an entire plant. By using symbols to represent various pieces of equipment, the flow diagram provides the designer with an overall view of the operation of a facility. The flow diagram is representative of the types used by many companies in the piping industry. While actual symbols may vary slightly from one company to the next, the "look and feel" of flow diagrams is the same throughout the chemical industry. Students must become familiar with the piping, equipment, instrumentation symbols and abbreviations used on flow diagrams, in order to be able to "read" and interpret them.

schematic *adj.* 图解的，概要的

abbreviation *n.* 缩写

interpret *v.* 解释，说明

simplistic *adj.* 过分简单化的

One of the most difficult concepts for students to comprehend is the absence of scale in the preparation of flow diagrams. The flow diagram should be laid out in a very simplistic and logical order and be read from left to right. It guides the drafter and designer in the same manner a road map guides a traveler.

M9-1
单词读音及例句

Notes:

One of the most difficult concepts for students to comprehend is the absence of scale in the preparation of flow diagrams.

译文：学生最难理解的观念之一是在绘制流程图时是缺乏比例的。

语法：句中 for students to comprehend 为不定式做后置定语，修饰 one of the most difficult concepts，students 为不定式结构的逻辑主语。

dedicate *v.* 致力，奉献

package *n.* 程序包

Process engineers are responsible for developing flow diagrams. In many large engineering firms, an entire department is dedicated to the development of flow diagrams. Today almost all flow diagrams are laid out with CAD, using third-party piping packages such as ProFlow or individually developed company packages.

M9-2
单词读音及例句

Flow diagrams belong to the documents of central importance for a process plant.

They are drawn up in three detail stages:

- basic flow diagram (BFD)
- process flow diagram (PFD)
- process and instrumentation diagram (P&ID)

2. Basic flow diagram

provision *n.* 规定

specification *n.* 规格

offsite *adj.* 厂区外的

depict *v.* 描写，描述

M9-3
单词读音及例句

While the basic diagrams and flow diagrams have to be drawn up within the framework of basic engineering, the editing of the detailed process and instrumentation diagrams (P & ID) occurs only after the order placing for the manufacturing of the plant within the framework of the detail engineering. Provisions regarding the generation of flow diagrams are either based on the customer specifications or agreed upon by operator and plant manufacturer. The basic flow diagram shows the main steps of a process. For clarity reasons, merely the main steps, plant lines and complete offsites are depicted as rectangles with plain text identification. The main mass flows are shown as simple lines with flow direction arrows.

Before attempting to calculate the material or energy requirements of a process it is desirable to attain a clear picture of the process. The best way to do this is to draw the aforementioned flow diagram where the flow diagram is defined as a line diagram showing the succession of steps in the process. Flow diagrams are very important for saving time and eliminating mistakes. The beginner should learn how to draw them properly and cultivate the habit of sketching them on the slightest excuse.

sketch *v.* 画草图

neat *adj.* 整齐的

elaborate *adj.* 精心制作的

duct *n.* 导管

chute *n.* 斜槽

M9-4
单词读音及例句

Notes:

The beginner should learn how to draw them properly and cultivate the habit of sketching them on the slightest excuse.

译文：初学者应该学习如何正确地画出它们（流程图），并养成有问题就画流程图的习惯。

语法：句中连接副词 how 引导宾语从句，从句到 excuse 结束，从句中 and 前后的 draw 和 cultivate 为并列结构。

The following rules should be observed.

- Show operating units by simple neat rectangles. Do not waste time in elaborating art work since it is without advantage or even meaning.
- Make each material stream line represent an actual stream of material passing along a pipe, duct, chute, belt, or other conveying device.

Refer to Figure 9-1(a). Show the gaseous mixture of carbon dioxide, oxygen, carbon monoxide, nitrogen, and water vapor obtained from the combustion of a hydrocarbon fuel and exiting from the stack of a furnace as a single material stream of flue gas, not to be confused as five separate streams. Figure 9-1(b) shows a sophisticated "device" that sorts components of the flue gas and delivers them as five separate products through five separate pipes.

gaseous *adj.* 气态的

combustion *n.* 燃烧

sophisticated *adj.* 复杂的

open steam 直接蒸汽

closed steam 间接蒸汽

coil *n.* 盘管

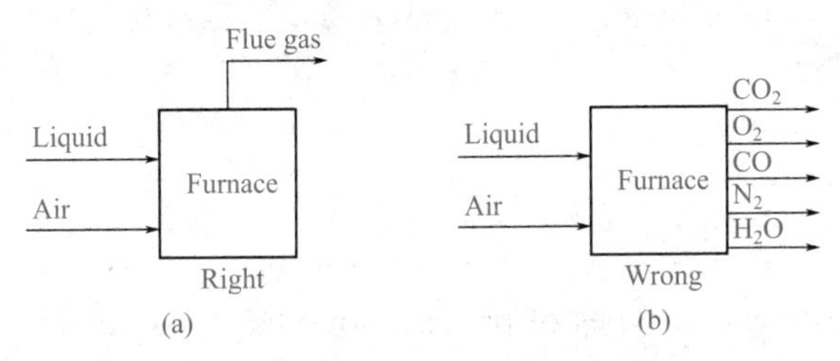

Figure 9-1　Right and wrong representation of a flue gas containing CO_2, O_2, CO, N_2, and water vapor

M9-5
单词读音及例句

- Distinguish between "open" and "closed" material streams. Open steam is being blown into the tank, mixing with the other materials in it. Closed steam is passing through a coil, being kept separate from the tank contents.

Notes:

Open steam is being blown into the tank, mixing with the other materials in it. Closed steam is passing through a coil, being kept separate from the tank contents.

译文：直接蒸汽被吹入罐中，与其中的其他物料混合。 间接蒸汽通过盘管流动，与罐内的物质保持分离状态。

语法：本句使用进行时强调两种蒸汽不同动作的持续性。此外句中 mixing 和 being kept 均为现在分词作状语。

- Distinguish between a continuous operation in which raw material is fed into the equipment in an uninterrupted stream, and a batch operation in which a fixed charge of material is introduced, processed, and removed, followed by the feeding of a new charge with repetition of the cycle, *etc.* Batch operation is indicated conveniently by putting a double bar on the material stream lines of entering materials (Figure 9-2).

Except for a few unusual situations which will be discussed later, keep flow diagrams free of data regarding the material streams. An over-burden of data clutters up the diagram and robs it of its main function, which is to present a clear picture of the materials as they move into, through, and out of the process. (From Charles Prochaska and Louis Theodore, Introduction to Mathematical Methods for Environmental Engineers and Scientists, 2018, 476）

over-burden 超负荷

clutter *v.* 使凌乱，杂乱

rob *v.* 抢夺

M9-6
单词读音及例句

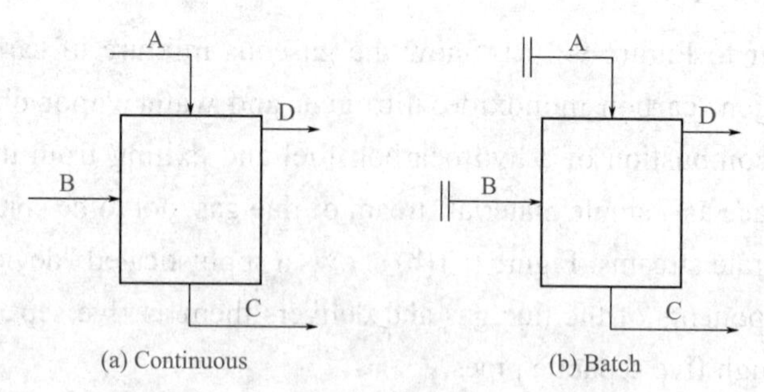

Figure 9-2　Continuous *vs.* batch operation

Reading comprehension

1. Please list several types of process flow diagrams commonly used in the chemical industry.

2. What are generally included in a basic flow diagram?

Exercise

1. True or False.

(1) Drawing a basic flow diagram requires elaborate art work.

(2) Each material stream line represents an actual stream of material passing along a pipe.

(3) Composition of the stream line should be indicated on a basic flow diagram.

2. Translate the following sentences into Chinese.

(1) The arrangement of the equipment reflects the logical sequence of flow depicted on the flow diagram.____________________________________

(2) The piping designer routes the pipe between two vessels as indicated by the flow diagram using piping specifications.________________________________

Further Reading

Simplified Flow Diagrams

In many cases simplified flow sheets are prepared to illustrate a process. These are often for a special purpose and do not show all the details of the process. A common type of flow sheet shows the major unit operations and chemical reactors with their interconnecting piping and an identification of the materials being processed. The units are not shown to scale, but the drawing may resemble the equipment used for the operation. Such a diagram would be called a graphic process flow sheet. For a small process of a few units, a pictorial flow sheet with drawings of equipment approximately to scale may be used. Photographs are usually unsatisfactory because the arrangements and connections are seldom clear. On a process flow sheet, the equipment is arranged logically to show the flow of materials through the process, but a photograph shows the final physical arrangement determined by structural requirements without regard to the process flow. All tall absorber columns may be grouped together for structural support, and large heat exchangers may be grouped together for ease of maintenance. The successful individual must learn to read flow sheets and to relate them to the actual plant layout. (From Charles Prochaska and Louis Theodore, Introduction to Mathematical Methods for Environmental Engineers and Scientists, 2018, 476）

pictorial *adj.* 绘画的，形象的

ease *n.* 容易

layout *n.* 空间布置

M9-7
单词读音及例句

Reading comprehension

1. Can you replace the flow diagram with actual plant photographs?
2. What is the difference between the flow diagram and the actual plant layout?

Lesson two: Process Flow Diagram

In this lesson you will learn:

- Process flow diagram
- Flow chart symbols
- Flow diagram instruments

1. Function of process flow diagram

The process flow diagram serves the depiction of the process concept and shows, compared to the basic flow diagram, the next higher detail stage. Usually it contains all essential components such as pumps, compressors, heat exchangers, columns, vessels *etc.*, the main pipelines and transport facilities as well as the most important measuring and control devices.

compressor *n.* 压缩机

denote *v.* 指示，表示

withstand *v.* 禁得住

commodity *n.* 物料，商品

M9-8
单词读音及例句

Notes:

The process flow diagram serves the depiction of the process concept and shows, compared to the basic flow diagram, the next higher detail stage.

译文：过程流程图服务于过程的描述，相较于基本流程图，它是更详细的展示。

语法：句中两逗号间插入的部分 compared to the basic flow diagram 为过去分词短语作状语。

The process flow diagram will denote the following:

- Conditions to be used for the design of various pieces of equipment (fractionation columns, pumps, heaters, *etc*.) required for facility operation.
- Operating and design conditions under which a particular unit or piece of equipment will normally operate. Design conditions establish the limits that equipment used in the facility can withstand. Design pressure is calculated to be at least 10% above the maximum operating pressure. The design temperature will be at least the maximum operating temperature, but should be at least 25 degrees above the normal operating temperature.
- Composition of the commodities used in the process sequence as they enter and leave the unit.

2. Flow chart symbols

Various symbols are universally employed to represent equipment, equipment parts, valves, piping, *etc*. Some of these are depicted in the schematic in Figure 9-3 as below.

Although a significant numbers of these symbols are used to describe some of the chemical and petrochemical processes, only a few are needed for simpler facilities. These symbols obviously reduce, and in some instances, replace detailed written descriptions of the process. Note that many of the symbols are pictorial, which helps in better describing process components, units and equipment.

description *n.* 说明书，说明

pictorial *adj.* 形象化的

sophistication *n.* 复杂

preparer *n.* 填表人，提出人

freehand *adj.* 手绘的

M9-9
单词读音及例句

Notes:

Note that many of the symbols are pictorial, which helps in better describing process components, units and equipment.

译文：请注意，许多符号都是形象化的，这有助于更好地描述过程组件、单元和设备。

语法：本句为祈使句；连词 that 引导宾语从句；关系代词 which 引导非限制性定语从句。

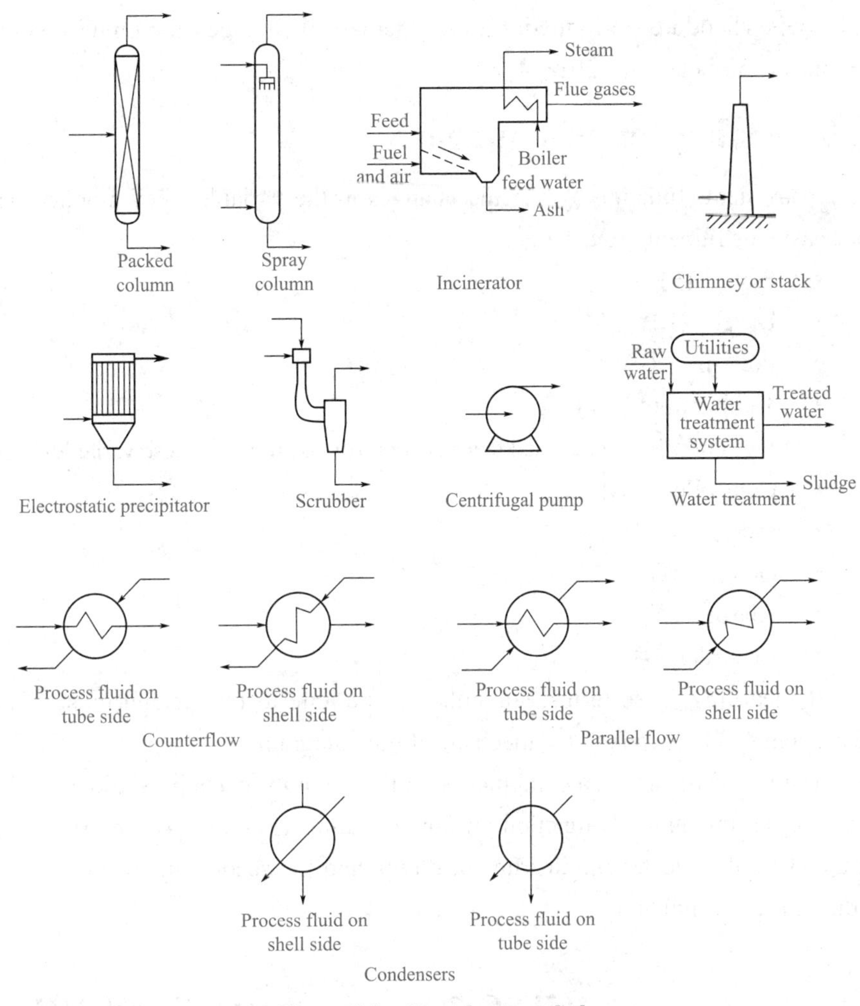

Figure 9-3 Flowchart symbols

The degree of sophistication and details of a flowchart usually varies with both the preparer and time. It may initially consist of a simple freehand block diagram with limited information that includes only the equipment; later versions may include line drawings with pertinent process data such as overall and componential flow rates, utility and energy requirements, environmental equipment, and instrumentation. During the later stages of a design project, the flowchart will be a highly-detailed P&I (piping and instrumentation) diagram; this aspect of the design procedure is beyond the scope of this text.

pertinent *adj.* 相关的

conceptually *adv.* 概念上

M9-10
单词读音及例句

In a sense, flowcharts are the international language of the engineer, particularly the practicing engineer. Chemical engineers conceptually view a (chemical) plant as consisting of a series of interrelated building blocks that are defined as units or unit operations. The plant ties together the various pieces of equipment that make up the process. Flow schematics follow the successive steps of a process by indicating where the pieces of equipment are located and the material streams entering and leaving each unit. (From Charles Prochaska and Louis Theodore, Introduction to Mathematical Methods for Environmental Engineers and Scientists, 2018, 476）

3. Flow diagram instruments

Instruments function by sensing changes in the variables they monitor. The four basic instrument groups are:

- Flow (F)
- Level (L)
- Pressure (P)
- Temperature (T)

The types of instruments used to sense, control and monitor these variables are:

- Controller (C)
- Indicator (I)
- Gauge (G)
- Alarm (A)
- Recorder (R)

indicator *n.* 指示器
gauge *n.* 计量器
alarm *n.* 警报器

M9-11
单词读音及例句

By learning these terms, students will be able to understand most of the instrument symbols found on a mechanical flow diagram.

The figure illustrates a combination of the symbols and abbreviations used to represent an instrument's function on flow diagrams (Figure 9-4). The first letter in the symbol indicates the instrument group, and the second and/or third letters indicate the instrument type.

65TI32
Local indicator on plant

65FRC27

65LC27
Device in control room mounted behind panel on plant

65LAH27
High-level alarm annunciator mounted on control panel

65 F RC 27

Plant section or unit number

First letter PROPERTY:
A analysis
D density
F flow
J power
L level
P pressure
R ratio
S speed
T temperature
V viscosity
W weight/force
Z position

Succeeding letters FUNCTION:
A alarm
C controller
E element
H high
I indicator
L low
R recorder
T transmitter
V valve
Y calculation

Loop number

Figure 9-4 Instrumentation tag bubble notation

To indicate a change or to control the flow, level, pressure, or temperature, an instrument must first sense a change in the variable. Once a change has been detected, the instrument then transmits this information *via* mechanical, electronic, or pneumatic means to a control panel where it can be observed and recorded. At the same time, the instrument may activate other devices to affect and change process conditions in the facility. Some instruments are read in the plant at the instrument's actual location. Others are displayed on a control panel located in an operator's control room.

facility　*n.* 设施，设备，场所

control panel 控制面板

M9-12
单词读音及例句

Notes:

Once a change has been detected, the instrument then transmits this information *via* mechanical, electronic, or pneumatic means to a control panel where it can be observed and recorded.

译文：一旦检测到变化，仪器就会通过机械、电动或气动方式将此信息传输到控制面板，在那里可以进行观察和记录。

语法：句首连词 once 引导状语从句，意为“一旦，一……就”；关系副词 where 引导定语从句，修饰 control panel。

(1) Gauges

Gauges are instruments that measure the liquid level inside a vessel or the temperature and/or pressure in the piping system. Level, temperature, or pressure gauges are locally mounted to enable plant operators to obtain a visual reading.

locally mounted 就地安装

regulate　*v.* 控制

M9-13
单词读音及例句

(2) Controllers

Devices used to maintain a specified liquid level, temperature, pressure, or flow inside a vessel or piping system. They activate the control valve that regulates the level, temperature, pressure, and flow in and out of the vessel.

(3) Alarms

Signals *via* lights or horns that indicate the liquid level, temperature, or pressure inside a vessel are too high or too low or that there is no flow or reverse flow.

(4) Indicators

Devices used to indicate the liquid level, temperature, pressure or flow rate inside a piping system.

(5) Recorders

Devices used to record the liquid level, temperature, pressure, and flow rate inside a vessel or piping system throughout a certain shift or period of time.

These same instruments may be found in combination such as Level Recording Controller. Here the instrument not only records the liquid level but also sends a signal to a control valve to control the liquid level inside the vessel. (From Roy A.Parisher and Robert A.Rhea, Pipe drafting and design, 2002, 308）

Reading comprehension

1. List five items shown on the process flow diagram.
2. List the four basic instrument groups.
3. What type of instrument is used to maintain a certain liquid level?

Exercise

1. Identify the following instrument abbreviations.

（1）LAH

（2）TRC

（3）PIC

2. Translate the following sentences into Chinese.

（1）There are several essential constituents to a detailed process flowchart beyond equipment symbols and process stream flow lines.________________

____

（2）"PD" would indicate the measurement of a pressure differential between two points in the process.________________________________

____

Further Reading

Divider Types of Instruments

The balloon dividers generally specify the "location" of instruments. What is important from an I&C(instrument and control) practitioner's viewpoint regarding "location" is if the instrument is in the field, in the control room, or in the field cabinets.

Different divider shapes are shown in Table 9-1 as below. I have shown the symbol with an irregular shape. This is so that you focus on the divider and not the shape.

For the divider, there are five different cases:

① No divider. This means that the instrument is outdoors, in the field. Examples are a flow element or sensor, or a level switch or gauge. They are connected to the control system, but they are not encased in a control room or in an auxiliary control cabinet. The majority of sensors are located outdoors and their tag doesn't have any divider.

② Single solid line. This shows that the instrument is situated inside the main control room. It also indicates that the instrument is accessible

practitioner *n.* 从业者

in the field 在现场，在野外

cabinet *n.* 机柜

auxiliary *adj.* 辅助的

tag *n.* 标识

M9-14
单词读音及例句

Table 9-1 Divider types

Items	Unenclosed location	Enclosed locations			
		In control room		In field cabinet	
	In field	In accessible location	In inaccessible location	In accessible location	In inaccessible location
Symbol (divider type)					
Examples	All sensors, all transmitters	Controllers, indicators	Non-important functions	Local alarms	Transducers

and visible to the operator.

③ Single dashed line. This shows that the instrument is located inside the main control room, but it is not that important and is inaccessible to the operator. In the past, before computerized control, instruments were mounted on a board in the control room. The important ones were mounted on the front of the board and those that were deemed less important were mounted on the back of the board. So, this is the same concept.

④ Double solid line. This indicates that the instrument is situated in a control cabinet in the field and is accessible and visible to the operator. Dedicated control cabinets are supplied by manufacturers, especially for PLC control of their equipment.

⑤ Double dashed line. This means that the instrument is in a field control cabinet, but inaccessible to the operator. (From Moe Toghraei, Piping and Instrumentation Diagram Development, 2019, 472）

situated *adj.* 位于

deem *v.* 认为，视为

dedicated *adj.* 专用的

M9-15
单词读音及例句

Reading comprehension

1. List five divider types of instrument symbols on the process flow diagram.

2. If a thermometer is situated inside the main control room, what kind of divider type should be used to specify the “location” of the instrument?

Lesson three: P&ID

In this lesson you will learn:

- P&ID
- How to show P&ID

1. What is P&ID

order place 下订单

parallel to 与……平行

procurement *n.* 采购

subdivide *v.* 细分

M9-16
单词读音及例句

Immediately after the order placing and parallel to the procurement of the components, the drawing up of the piping and instrumentation diagrams (P&ID) is started. A great deal of the plant manufacturer's know-how will find its way into the P&I diagrams, as they contain all essential information regarding the designed plant in the form of a code.

The information given in the P&I diagrams is subdivided into basic information which has to be included at any rate and additional information which may be provided.

Notes:

The information given in the P & I diagrams is subdivided into basic information which has to be included at any rate and additional information which may be provided.

译文：P&ID 图中给出的信息被细分为无论如何都必须包含的基本信息和可能提供的附加信息。

语法：at any rate 意为“不管怎样，无论如何，至少”；given… diagrams 为过去分词短语作后置定语；句中两个 which 均为关系代词，引导定语从句。

mould *v.* 模具浇铸

tee *n.* T形管件

nominal width 公称宽度

nominal pressure 公称压力

insulation *n.* 绝缘

gradient *n.* 梯度

(1) Basic information

- All appliances and machinery, including motors.
- All pipes or means of transport (*e.g.* conveyor belts).
- All pipe installations (*e.g.* valves and fittings) and moulded parts (*e.g.* tees).
- Indication of nominal widths, nominal pressures and piping class.
- Indication regarding insulation, electric pipe heater as well as gradients.
- All measuring and control equipment with the respective action lines.
- Identification of all components according to a standard numbering system.
- Indication of the drawing numbers for the material flow entering and leaving the P&ID.

(2) Additional information

centrifugal pump 离心泵

appliance *n.* 器具，设备

elevation *n.* 标高，高度

Operating data of appliances (*e.g.* pump head and flow rate in the design point of a centrifugal pump).

- Material flow numbers or material flow bars.
- Indication of appliance material.
- Elevation of the main appliances.

M9-17
单词读音及例句

M9-18
单词读音及例句

- Other comments.

The kind of depiction is as follows:

- Drawing according to HG/T 20519—2009.
- Regarding their elevation and size, appliances and machinery have to be depicted roughly to scale.
- Regarding their function, pipes, valves and equipments have to be depicted roughly according to their position.

> Notes:
>
> Regarding their elevation and size, appliances and machinery have to be depicted roughly to scale.
>
> 译文：关于它们的高度和尺寸，设备和机械必须按比例粗略地描绘。
>
> 语法：to scale 意为“按一定比例”，be depicted roughly to scale 意为“按照比例大致描述”。

When drawing up the process flowsheets it is often not yet clear which type of pump or fitting or which type of agitator will finally be applied. Therefore, the standards mentioned above contain general flowsheet symbols.

In the P&IDs the design descriptions and their corresponding symbols have to be inserted. It has also to be emphasized that all components and aggregates of the designed plant must be included in the P&IDs. The demand for such completeness results, inter alia, from the fact that the P&IDs are the basis for the compilation of lists and that the P&IDs are used for the later test on completion at the end of the assembly. (From Frank Peter Helmus, Process Plant Design, 2008, 204）

agitator *n.* 搅拌器

emphasize *v.* 强调，着重

aggregate *n.* 集合体，组合体

M9-19 单词读音及例句

2. How to show P&ID

The first thing that a reader of a P&ID expects is legibility. A P&ID set is developed by the engineers during the design step of the project but will later be used in a process plant during operations. It is used by individuals with different levels of knowledge on and familiarity with a P&ID. A P&ID may need to be read by engineers, managers with management degrees, trade practitioners, and many others. It also should give enough information about the project that readers from different backgrounds will be able to understand it. A P&ID can be used during normal operations or an emergency event. Therefore, a P&ID should be as legible as possible. Below are some rules of thumb regarding the visual aspects of P&IDs:

- The P&ID sheet is almost always in landscape orientation.
- Limit the number of main equipment shown on each sheet. A crowded P&ID should be avoided. Some companies try to limit the number of items on each P&ID to four or five, but other companies allow as many as 10.

inter alia 尤其是

legibility *n.* 易读性，易辨认

rules of thumb 经验法则

landscape *n.* 横向打印（文件）

orientation *n.* 方向

M9-20 单词读音及例句

- A P&ID is a pictorial document, so minimize notes on the P&IDs (no "note-stuffed" drawings).
- Draw P&ID symbols as similar as possible to what an item looks like in reality and approximate to relative size, but remember, a P&ID is not drawn to scale.
- Do not represent the real length of pipes on P&IDs. P&ID is a "Not To Scale" (NTS) drawing. Therefore, a short line on a P&ID could represent a few hundred meters of pipe.
- Do not "pack" symbols on one side of the P&ID sheet. The symbols should be fairly spread out (Figure 9-5).

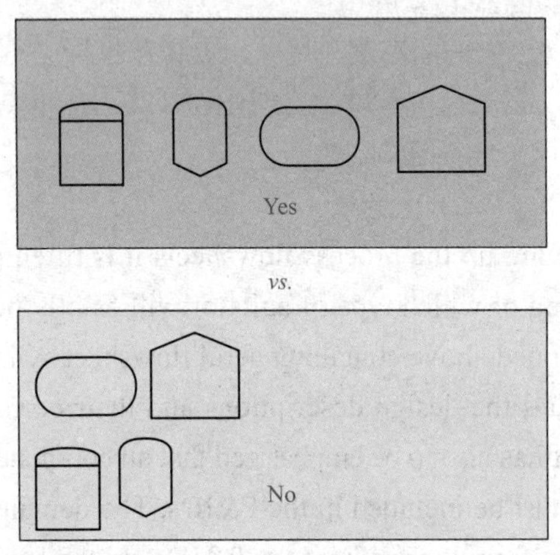

Figure 9-5 Equipment should be fairly distributed horizontally on the P&ID

horizontally *adv.* 水平地
stack *v.* 堆放
callout *n.* 图例，零件编号

- Generally, an equipment is arranged horizontally on one level. On an ideal P&ID sheet, any imaginary vertical line should, at maximum, cross one equipment symbol. The P&ID will be too difficult to understand when equipment symbols are stacked on each other and the connection between the symbols and the equipment callouts are difficult to be identified.

M9-21
单词读音及例句

Notes:

The P&ID will be too difficult to understand when equipment symbols are stacked on each other and the connection between the symbols and the equipment callouts are difficult to be identified.

译文：当设备符号相互堆叠且符号与设备标注之间的连接难以识别时，P&ID 将难以理解。

语法：句中的 too… to… 结构意为“太……而不能”；连词 when 引导状语从句。

- All the different elements on a P&ID sheet should be connected to each other. If a group of elements have no connection to the rest of the elements, they can be drawn on another P&ID sheet.

- No visiting stream is allowed. A visiting stream is a stream that has no connection with other items on the P&ID.
- Do not try to present a P&ID in a way that follows geographical directions (*e.g.* north). P&IDs are drawings independent of these.

(1) Line crossing over

Lines (in different forms) are symbols for pipes or signals. Crossing lines should be avoided or kept to a minimum, as well as changing the direction of lines. However, sometimes both are inevitable, and it is common to see lines crossing each other in P&IDs. It is important to know that crossing lines in a P&ID does not reflect the reality of pipe routes in field. Crossing lines in the P&IDs is not acceptable aesthetically, but it is unavoidable in some cases. However, pipecrossing (or better, pipe clashing) in field shows a mistake in the design. A crossover could be shown in P&IDs in two forms, a "jump" or a "jog" (Figure 9-6).

inevitable　*adj.* 不可避免的

aesthetically　*adv.* 审美地

intact　*adj.* 原封不动的，完整的

prevailing　*adj.* 流行的

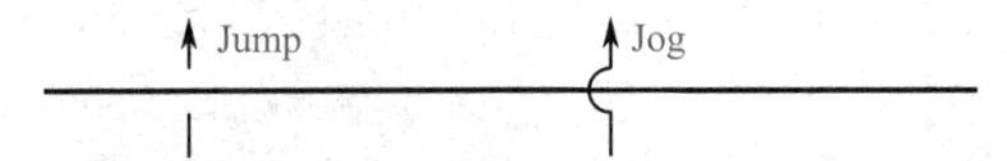

Figure 9-6　P&ID representation of pipe crossing

M9-22
单词读音及例句

The decision on which line should be "manipulated" and which remains intact will be based on the prevailing line rule (Horizontal process lines > Vertical process lines > Instrument signals).

For example (Figure 9-7), in crossing a horizontal process line and a vertical process line, the horizontal process line remains intact.

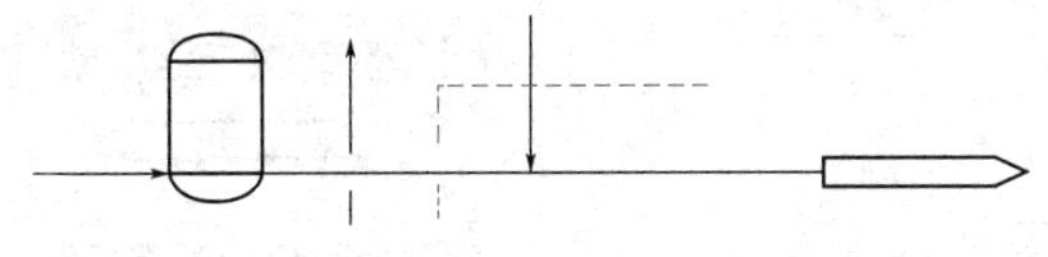

Figure 9-7　Examples of line crossing

(2) Equipment crossing

Equipment-equipment crossing is not allowed on a P&ID. Meanwhile, equipment-line crossing can be shown, but it is not a good practice. This can be managed by using breaks on the line (Figure 9-8).

off-page connector 离页连接符

exception　*n.* 例外

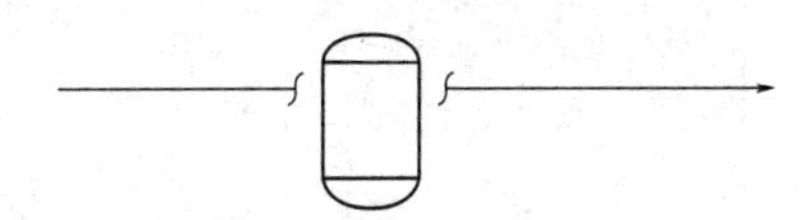

Figure 9-8　Equipment-line crossing is allowed

(3) Off-page connector

Off-page connectors (Figure 9-9) are continuation indicators for the lines, which are pipe and signal symbols. The preference is to show off-page connectors horizontally and at the edge of a P&ID sheet. The incoming off-page connectors are preferably located at the left edge of the P&ID sheet, and the outgoing

M9-23
单词读音及例句

off-page connectors at the right edge of the P&ID sheet. There are, however, exceptions. The off-page connectors in network P&IDs can be drawn in vertical positions, too. This is because network P&IDs should generally follow the plot plans, and sometimes using vertical off-page connectors prevents crowding on the P&ID. Some companies adopted this practice to decrease crowding on the P&ID by avoiding drawing of long utility lines, which are not as important as process lines. The incoming off-page connectors can be on the right edge of the P&ID sheet if they are coming from a downstream of the process. The outgoing off-page connectors could be on the left edge of the P&ID sheet if they are going somewhere on the upstream of the process. (From Moe Toghraei, Piping and Instrumentation Diagram Development, 2019, 472)

Notes:

Some companies adopted this practice to decrease crowding on the P&ID by avoiding drawing of long utility lines, which are not as important as process lines.

译文：一些公司采用避免绘制长的公用工程线的做法，从而减轻P&ID图上的拥挤，这些公用工程线不如流程线重要。

语法：关系代词 which 引导非限制性定语从句，修饰 long utility lines。

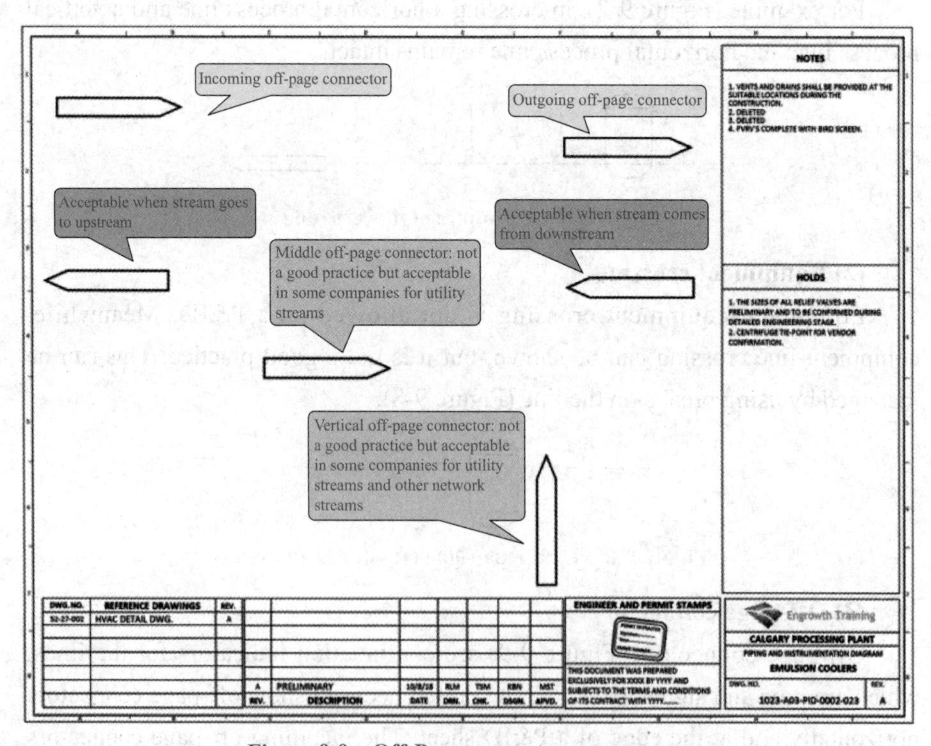

Figure 9-9 Off-Page connector appearance

Reading comprehension

1. What is a P&ID?
2. Why is P&ID important?
3. Which line should be “manipulated” and which remains intact on a P&ID?

Exercise

1. True or False.

（1）The P&ID sheet is always in landscape orientation.

（2）A P&ID should be drawn to scale.

（3）Visiting stream on a P&ID is allowed.

（4）Equipment-line crossing on a P&ID is allowed.

2. Translate the following sentences into Chinese.

（1）It is important to know that a P&ID will be used in the process plant during operation, including emergency situations.______________________________

____

（2）A straight piece of pipe in-field can be represented as a line with several directional changes on a P&ID.______________________________

____

Further Reading

PFD and P&ID

To ensure a certain transparency of the total plant system, the database is subdivided into components or systems. Here it is advisable to choose the systems in such a way that they correspond to the later system or component controls. The starting point for it is the process flow diagram. Usually, each main component represents an individual system. For each system, an individual P&I diagram can be drawn up. The systems can then again be subdivided *e.g.* into groups or subgroups.

As the process flow diagrams (PFD) only contain the main components, fittings, pipes and measuring points, the P&IDs have now to be completed. Some examples for elements not included in the process flowsheets are:

- Drain connections, fittings and pipes.
- Flushing connections, fittings and pipes.
- Exhaust connections, fittings and pipes.
- Pipework installations such as steam traps, blanks, orifice plates,

transparency *n.* 透明度

flush *v.* 冲洗

exhaust *n.* 废气，排气管

steam trap 疏水阀

blank *n.* 盲板

M9-24
单词读音及例句

expansion joint 膨胀节
seal *n.* 密封
burst disk 安全隔板
coupling *n.* 接头
accessory *n.* 附件，配件
sump *n.* 水坑，井
dosing *n.* 定量给料

M9-25
单词读音及例句

expansion joints, manual fittings, double shut-off device *etc*.

- The complete water system for the flushing of the mechanical seals.
- Sampling points.
- Safety installations like eye washes, safety showers, safety valves, burst disks, safety-relevant measuring points (*e.g.* safety temperature or pressure limiting valves).
- Maintenance equipment: compressed air and flushing water connections with hose couplings.
- Binary measuring instruments (*e.g.* leakage detector).
- Local measurement instruments (*e.g.* on-site manometer or thermometer).
- Other equipment accessories: horns, sirens, roof cranes, pump sump, ground inlets *etc*.

As already mentioned, small plants are integrated in the larger plant (*e.g.* a dosing plant with smaller vessels, complete internal piping and fittings as well as the related measuring instruments). In the P&ID, this can be depicted as a so-called black box. Then only the interfaces have to be connected to the black box.

Alternatively, the flowsheet drawn up by the relevant subcontractor may be adopted for the P&ID. In all probability the plant to be designed will include atmospheric vessels, simple or automated pump units, heat exchangers and dosing plants. (From Frank Peter Helmus, Process Plant Design, 2008, 204）

Reading comprehension

1. What is the connection between PFD and P&ID ?
2. List five safety installations that should be completed on a P&ID.

项目四
单元操作与控制

Part Ⅳ:
Unit Operation and Control

第十单元　动量传递及其应用
Unit 10　Momentum Transfer and Its Application

Lesson one: Fluid Flow

In this lesson you will learn:

- Fluid flow phenomena
- Basic equations related to fluid flow
- Centrifugal pump

1. Fluid flow phenomena

A fluid may be defined as a substance that does not permanently resist distortion. Density and viscosity are the most important physical and transport properties in fluid flow studies. These properties influence the power requirements during fluid transport and flow characteristics within the pipeline. A better understanding of these properties helps process engineers design an optimum flow transport system.

The density of fluid is defined as mass per unit volume (kg/m^3). The density of fluid is a function of temperature and pressure. If the change in density is small with moderate temperature and pressure, the fluid is termed "incompressible"; if the changes are noteworthy, the fluid is termed "compressible". Liquids are generally incompressible whereas gases are compressible. The density of water is maximum at 4℃ and decreases with increase in temperature.

distortion　*n.* 变形
viscosity　*n.* 黏度
density　*n.* 密度
optimum　*adj.* 最佳的
noteworthy　*adj.* 显著的

M10-1
单词读音及例句

Notes:

If the change in density is small with moderate temperature and pressure, the fluid is termed "incompressible"; if the changes are noteworthy, the fluid is termed "compressible".

译文：如果在温和的温度和压力下密度变化很小，则流体称为"不可压缩流体"；如果变化较大，则流体被称为"可压缩流体"。

语法：分号前后是两个独立的句子，在英语中，分号是用于连接两个有关联，但同时都是独立句子的标点符号。句中两个 if 为连词，引导条件状语从句。

inherent *adj.* 固有的，内在的

adjacent *adj.* 相邻的，临近的

M10-2 单词读音及例句

sandwiched *v.* 被夹在中间的

plane *n.* 平面，水平

M10-3 单词读音及例句

A fluid may be assumed to be matter composed of different layers. A fluid starts to move when an external force is applied. Viscosity is an inherent property that resists the relative movement of adjacent layers in the fluid. The viscosity of a fluid usually varies significantly with temperature and remains independent of pressure. For most fluids, as the temperature of fluid increases, the viscosity decreases.

Figure 10-1 illustrates a thin layer of fluid sandwiched between two parallel plates, of area *A*, separated by a small distance *h*, the bottom one fixed and the top one subject to an applied force parallel to the plate, which is free to move in its plane. The fluid is considered to adhere to the plates and its properties can be classified by the way the top plate responds when the force is applied.

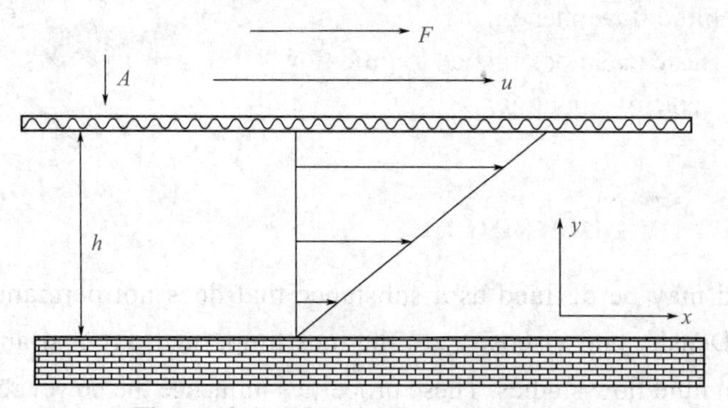

Figure 10-1　Material flow through pipe

Notes:

The fluid is considered to adhere to the plates and its properties can be classified by the way the top plate responds when the force is applied.

译文：流体被认为粘贴在板上，其特性可以根据施加外力时上板的响应方式进行分类。

语法：句中 the top plate responds 为定语从句，修饰 the way，引导词 that 被省略。

Application of a force F to the upper plate causes it to move at velocity u. The fluid continues to deform as long as the force is applied, unlike a solid, which would undergo only a finite deformation. For an incompressible, the resulting shear stress (τ) is equal to the product of the shear rate (velocity gradient du/dy) and the viscosity μ of the fluid medium. Thc equation which describes the behavior is:

$$\tau = \mu \cdot du/dy$$

Newtonian fluids exhibit a straight line relationship between shear stress and shear rate with zero intercept.

velocity　*n.* 流速

deform　*v.* 变形

Newtonian fluids 牛顿流体

conservation　*n.* 保持，守恒

M10-4
单词读音及例句

2. Basic equations related to fluid flow

(1) Equation of continuity

The principle of the conservation of matter is used to handle fluid flow problems. The equation of continuity states that for an incompressible fluid flowing in a tube of varying cross-section, the mass flow rate is the same everywhere in the tube (Figure 10-2). The mass flow rate is simply the rate at which mass flows past a given point, *i.e.* total mass flowing past divided by the time interval.

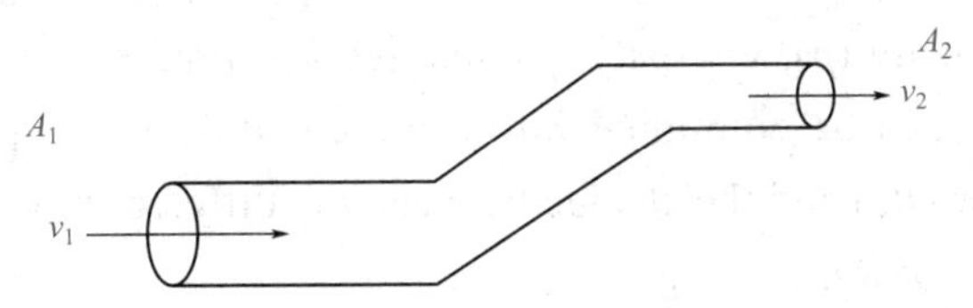

Figure 10-2　Fluid flow through varying cross-section of pipe

The equation of continuity for location 1 and 2 can be expressed as:

$$\rho_1 A_1 v_1 = \rho_2 A_2 v_2$$

where ρ is the density (kg/m^3), v is the velocity (m/s^1) and A is the area (m^2). Generally, the density of the fluid remains constant and thus the equation changes to the following form:

$$A_1 v_1 = A_2 v_2$$

(2) Reynolds number

Fluid flowing in pipes exhibits two primary flow patterns. At low velocities fluids tend to flow without lateral mixing and adjacent layers slide past one another. Reynolds demonstrated this by injecting a thin stream of dye into a fluid and finding that it ran in a smooth stream in the direction of flow. Such straight-line type flow is known as laminar flow. In the same experiment, Reynolds found that as the velocity of flow increased, the smooth line of dye changed until finally, at high velocities, the dye was rapidly mixed into the disturbed flow of the surrounding fluid. Such flow is known as turbulent flow. There is also a critical zone when the flow can be either laminar or turbulent. This intermediate flow is termed transitional flow. This instability in fluid behavior passing through a

lateral　*adj.* 横向的

Reynolds number 雷诺数

laminar flow　层流

turbulent flow 湍流

transitional flow 过渡流

M10-5
单词读音及例句

pipe can be predicted in terms of the relative magnitudes of the velocity and the viscous forces acting on the fluid.

> Notes:
>
> This instability in fluid behavior passing through a pipe can be predicted in terms of the relative magnitudes of the velocity and the viscous forces acting on the fluid.
>
> 译文：通过管道的流体表现的这种不稳定性可以根据流速和作用在流体上的黏性力的相对大小来预测。
>
> 语法：句中 passing… pipe 和 acting… fluid 均为现在分词短语作后置定语。

quantity *n.* 量

dimensionless *adj.* 无量纲的，无因次的

M10-6
单词读音及例句

The kinetic forces have a tendency to maintain the flow in its general direction, whereas the viscous forces tend to retard this motion and to preserve order and reduce eddies. The transition between laminar and turbulent flow appears when fluid velocity exceeds a critical limit. Reynolds studied this transition and found that the phenomenon depends on four quantities: pipe diameter (D), density (ρ), viscosity (μ) and average liquid velocity ($\bar{V}$). These four parameters can be combined into a dimensionless group known as the Reynolds number (Re) and the change between the different types of flow occurs at a definite value of Re:

$$Re = \frac{D\bar{V}\rho}{\mu}$$

The flow inside a pipe is laminar at Re below 2100, while flow becomes turbulent when Re exceeds about 4000. The flow is considered in transition when Re varies between 2100 and 4000.

Bernoulli equation 伯努利方程

(3) Bernoulli equation

The principle of the conservation of energy states that in a steady flow the sum of all forms of mechanical energy in a fluid along a stream line is the same at all points on that stream line.

> Notes:
>
> The principle of the conservation of energy states that in a steady flow the sum of all forms of mechanical energy in a fluid along a stream line is the same at all points on that stream line.
>
> 译文：能量守恒原理指出，在稳定流动中，沿流线的流体中所有形式的机械能的总和在该流线上的所有点都是相同的。
>
> 语法：句中第一个 that 为连词，引导宾语从句，宾语从句为主系表结构，主干为 the sum is the same，其余部分为修饰语。

This requires that the sum of kinetic energy and potential energy remains constant. Let us consider a parcel of fluid moving along the stream line in a pipe from location 1 to location 2 as illustrated in Figure 10-3. The flow is considered as steady, inviscid (viscosity of the fluid is zero) and the liquid is incompressible.

parcel *n.* 小包，包裹

inviscid *adj.* 非黏性的

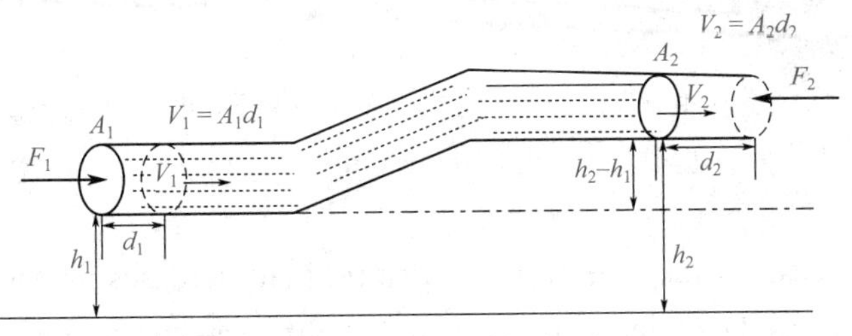

Figure 10-3 Bernoulli equation

$$\frac{P_1}{\rho} + \frac{1}{2}v_1^2 + gh_1 = \frac{P_2}{\rho} + \frac{1}{2}v_2^2 + gh_2 \tag{10-1}$$

Eq. (10-1) is known as the Bernoulli equation for an incompressible inviscid fluid without friction. The equation is one of the foundations of fluid mechanics. It is a mathematical expression, for fluid flow, of the principle of the conservation of energy and it covers many situations of practical importance. Each term in the above equation is a scalar and has the dimension of energy per unit mass.

friction *n.* 摩擦力

scalar *n.* 数量，标量

centrifugal pump 离心泵

discharge head 输送压头

3. Centrifugal pump

A centrifugal pump is one of the most commonly used pumps in the chemical processing industries for pumping water and Newtonian fluids. These pumps are available in a wide range of sizes, in capacities from 0.5 m^3/h to 2×10^4 m^3/h, and for discharge heads (pressures) from a few meters to approximately 48 MPa. The size and type best suited to a particular application can be determined only by an engineering study of the problem. A centrifugal pump consists of two basic parts, a rotating part and a stationary part. The rotating part includes the impeller and shaft, and the stationary part includes the casing, casing cover and bearing. The impeller contains a number of blades, called vanes, which are usually curved backward. There are three general types of casings: circular, volute, and diffuser. Each type consists of a chamber in which the impeller rotates, and an inlet and exit for the liquid being pumped. A circular casing contains an annular chamber around the impeller.

impeller *n.* 叶轮

shaft *n.* 传动轴

casing *n.* 机壳

bearing *n.* 轴承

vane *n.* 叶片

volute *n.* 螺旋形，蜗壳

chamber *n.* 室，房间

Notes:

Each type consists of a chamber in which the impeller rotates, and an inlet and exit for the liquid being pumped.

译文：每种类型都包含一个叶轮在其中旋转的腔室，以及一个用于泵送液体的入口和出口。

语法：句中 in which 引导定语从句，修饰 a chamber；being pumped 为现在分词作后置定语，修饰 the liquid。

spiral　*n.* 螺旋形

cross-sectional　*adj.* 横截面的

discharge　*n.* 排出

suction nozzle　吸嘴

cavity　*n.* 空腔

tangential　*adj.* 切线的，正切的

radial　*adj.* 径向的

M10-10
单词读音及例句

Volute casings take the form of a spiral that increases uniformly in cross-sectional area as the outlet is approached. In a diffuser-type casing, guide vanes or diffusers are interposed between the impeller discharge and the casing chamber. Losses are kept to a minimum in a well-designed pump of this type, and improved efficiency is obtained over a wider range of capacities.

External power is supplied to the shaft, rotating the impeller within the stationary casing. The liquid enters the suction nozzle and then enters the eye (center) of a revolving device known as the impeller. When the impeller rotates, it moves the liquid sitting in the cavities between the vanes outward and provides centrifugal acceleration. The velocity head it has acquired when it leaves the blade tips is changed to pressure head as the liquid passes into the volute chamber and then out through the discharge. Because the impeller blades are curved, the fluid is pushed in a tangential and radial direction by centrifugal force. The impeller can have 4-24 blades.

Notes:

The velocity head it has acquired when it leaves the blade tips is changed to pressure head as the liquid passes into the volute chamber and then out through the discharge.

译文：当液体进入蜗壳然后通过排放口排出时，它离开叶片尖端时获得的速度头变成压力头。

语法：句中 it has acquired 为定语从句，修饰 the velocity head，引导词 that 被省略；when 为连词，引导时间状语从句。

kinetic energy　动能

impart　*v.* 给予

M10-11
单词读音及例句

Figure 10-4 illustrates a cross-section of a centrifugal pump indicating the movement of liquid.

The key point is that the energy created by the centrifugal force is kinetic energy. The amount of energy supplied to the liquid is proportional to the velocity at the edge or vane tip of the impeller. The velocity of the liquid at the vane tip will be higher as the impeller rotates faster or with a large impeller and thus the greater the energy imparted to the liquid. (From Jasim Ahmed and Mohammad Shafiur Rahman, Handbook of Food Process Design, 2012, 1535）

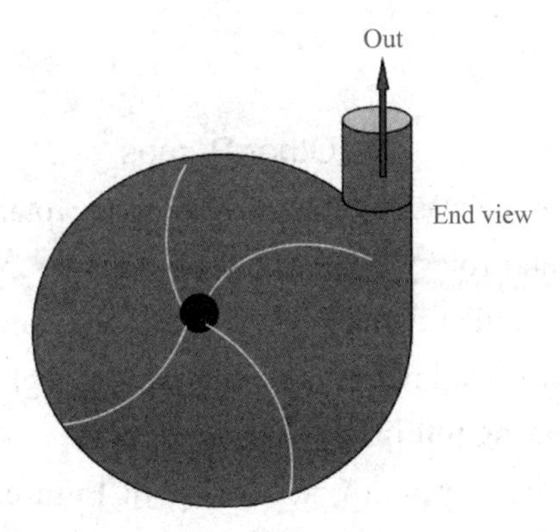

Figure 10-4　Centrifugal pump

Reading comprehension

1. What are important physical properties in fluid flow phenomena?

2. Could Bernoulli equation be used for conservation of energy states of gas transportation process?

3. What are the main structural components of centrifugal pumps?

Exercise

1. True or False.

（1）Priming is required before starting a centrifugal pump in order to remove all the air contained in the pump.

（2）The velocity of the liquid at the vane tip relates to the rotation velocity and dimension of impeller.

（3）At low velocities fluids tend to flow without lateral mixing, and such flow is known as turbulent flow.

2. Translate the following sentences into English or Chinese.

（1）In addition, the centrifugal pump normally cannot transfer liquid having vapor content.____

____

（2）The kinetic energy of the fluid increases from the center of the impeller to the tips of the impeller vanes.____

____

（3）使用离心泵时的管道布置通常也很简单。____________________________

____

reciprocating pump　往复泵

rotary　*adj.* 旋转式的

displacement pump　容积泵

M10-12
单词读音及例句

piston　*n.* 活塞

suction stroke　吸气冲程

expel　*v.* 排出

intermittent　*adj.* 间歇的

M10-13
单词读音及例句

Further Reading

Other Pumps

Pumps may be classified as reciprocating, rotary, or centrifugal. The reciprocating and rotary types are referred to as positive displacement pumps because unlike the centrifugal type, the liquid or semiliquid flow is broken up into small portions as it passes through the pump.

Reciprocating pump

A reciprocating pump, as shown in Figure 10-5, consists of a cylinder in which a piston moves, suction and delivery pipes, suction and delivery valves, and a rotating device that moves the piston. When the piston moves backward, the volume of the cavity increases and pressure inside it drops. This vacuum causes the suction valve to open and fluid enters the cylinder until the differential pressure between the supply side and the chamber is zero. This is called the suction stroke. When the piston moves forward, the volume of the chamber decreases which increases the pressure inside the chamber. Because of the increase in pressure, the inlet valve closes and the outlet valve opens and this expels the fluid from the chamber to the discharge line until the differential pressure is zero. This is known as the delivery stroke. This cycle repeats and intermittent discharge of fluid is obtained.

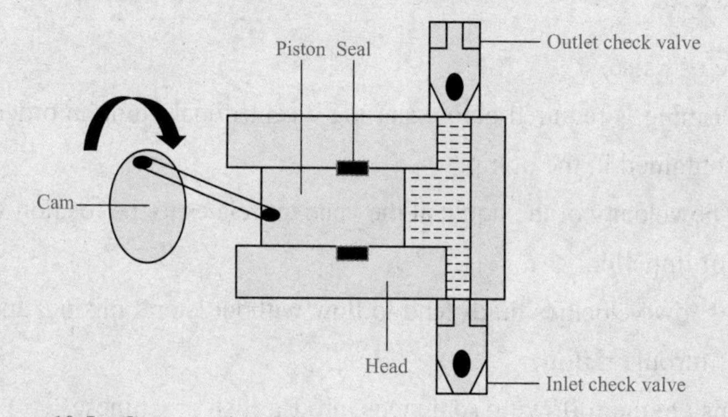

Figure 10-5　Cross-sectional diagram of a simple single-piston reciprocating pump

Notes:

This vacuum causes the suction valve to open and fluid enters the cylinder until the differential pressure between the supply side and the chamber is zero.

译文：这种真空导致吸入阀打开，流体进入气缸，直到供应侧和腔室之间的压差为零。

语法：句中 to open 为不定式作宾语补足语，这个结构中的谓语为 causes，宾语为 the suction。

Rotary pump

In general, rotary pumps operate in a circular motion and displace a constant amount of liquid with each revolution of the pump shaft. Howcver, a wide range of pumps occupy this category and their construction and operating principles are different. An external gear pump, an example of a rotary pump, is shown in Figure 10-6.

revolution *n.* 旋转
gear *n.* 齿轮

M10-14
单词读音及例句

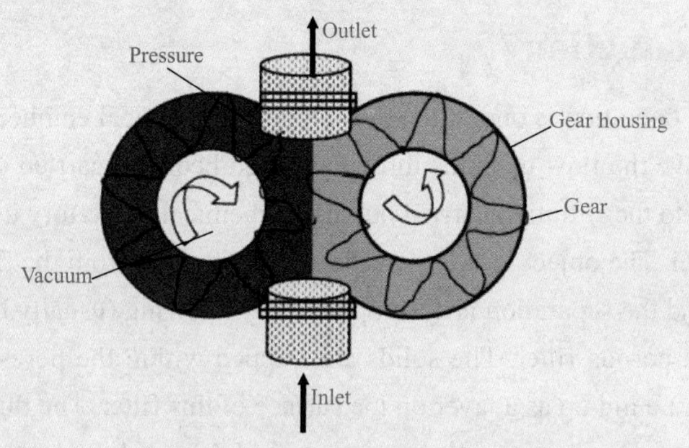

Figure 10-6 An external gear pump

An external gear pump consists of two identical gears rotating against each other. As the gears rotate, the teeth disengage and create an expanding volume on the inlet side of the pump. Fluid flows into the cavity, becomes trapped by the housing, and travels until it reaches the discharge side of the pump. Finally, the meshing of the gears forces liquid through the outlet port under pressure. In a rotary gear pump, the pump size and shaft rotation speed determine how much water is pumped per hour.

In general, priming is required before starting a centrifugal pump in order to remove all the air contained in the pump. Priming is carried out by filling the pump fully with the liquid to be pumped. Some self-priming pumps, such as reciprocating pumps of piston or plunger type, do not require priming. Positive displacement pumps of the rotating type, such as rotary or screw pumps, are also self-priming. These types of pumps have clearances that allow the liquid in the pump to drain back to the suction. (From Jasim Ahmed and Mohammad Shafiur Rahman, Handbook of Food Process Design, 2012, 1535)

disengage *v.* 脱离
housing *n.* 外壳
meshing *n.* 啮合
priming *n.* 灌泵
clearance *n.* 间隙

M10-15
单词读音及例句

Reading comprehension

1. Can a reciprocating pump deliver fluid continuously?

2. Is it necessary to prime the pump before starting a reciprocating pump or rotary pump?

Lesson two: Filtration

In this lesson you will learn:

- Filtration
- Filtration equipments

1. Introduction

funnel　*n.* 漏斗

porous filter
多孔过滤器

M10-16
单词读音及例句

This operation is one of the most common chemical engineering applications that involve the flow of fluids through packed beds. As carried out industrially, it is similar to the filtration carried out in the chemical laboratory using a filter paper in a funnel. The object is still the separation of a solid from the fluid in which it is carried and the separation is accomplished by allowing (usually by force) the fluid through a porous filter. The solids are trapped within the pores of the filter and (primarily) build up as a layer on the surface of this filter. The fluid, which may be either gas or liquid, passes through the bed of the solids and through the retaining filter. (From J. Patrick Abulencia and Louis Theodore, Open-Ended Problems 'A Future Chemical Engineering Education Approach', 2015, 583)

Notes:

The object is still the separation of a solid from the fluid in which it is carried and the separation is accomplished by allowing (usually by force) the fluid through a porous filter.

译文：（过滤的）目标仍然是将固体与输送它的流体分离，分离是通过让（通常是用外力）流体通过多孔过滤器来实现的。

语法：in which 引导定语从句，修饰 the fluid，从句在 and 前结束，and 后为并列分句。

suspension　*n.*
悬浮液

sludge　*n.* 烂泥

fine　*adj.* 细小的

M10-17
单词读音及例句

Filtration may therefore be viewed as an operation in which a heterogeneous mixture of a fluid and solid particles is separated by a filter medium that permits the flow of the fluid but retains the particles of the solid. Therefore, it primarily involves the flow of fluids through porous media.

Vacuum filtration finds wide application in the partial separation of liquids from concentrated suspensions, sludges, and slurries. When the liquid phase is highly viscous, or when the solids are so fine that vacuum filtration is too slow, pressure filtration provides a convenient solution to the separation problem. Centrifugal filtration is used when the solids are easy to filter and a filter cake of low moisture content is desired.

In all filtration processes, the mixture, or slurry, flows as a result of some driving force, *e.g.*, gravity, pressure (or vacuum), or centrifugal force. In each case, the filter medium supports the particles as they form a porous cake. This cake, supported by the filter medium, retains the solid particles in the slurry with successive layers added to the cake as the filtrate passes through the cake and medium.

Notes:

This cake, supported by the filter medium, retains the solid particles in the slurry with successive layers added to the cake as the filtrate passes through the cake and medium.

译文：这个由过滤介质支撑的滤饼，在滤液通过滤饼和介质时将悬浮液中的固体颗粒留下来，形成连续的滤层添加到滤饼中。

语法：句中 supported... 为插入的过去分词短语作定语，修饰 this cake；added to the cake 为过去分词短语，构成 with+n.+ 过去分词的复合宾语结构；as 为连词，引导状语从句。

2. Filtration equipment

The several methods for creating the driving force on the fluid, the different methods of cake deposition and removal, and the different means for removal of the filtrate from the cake subsequent to its formation, result in a great variety of filter equipment. Filters remove particles from a fluid stream using porous materials such as wire mesh, solids beds, cloth, and sintered metal (barrier). The porous medium catches the particles and allows the fluid to pass through. A pressure difference is always created across the porous barrier. The effluent from the barrier is known as the filtrate. In general, filters may be classified according to the nature of the driving force supporting filtration.

deposition *n.* 沉积
wire mesh 金属网
sintered *adj.* 烧结的
filtrate *n.* 滤液

(1) Rotary vacuum filters

Rotary vacuum filters are used where a continuous operation is desirable, particularly for large-scale municipal solids dewatering operations. The filter drum is immersed in a slurry where a vacuum is applied to the filter medium that causes the cake to deposit on the outer surface of the drum as it rotates through the slurry.

rotary vacuum filter 转鼓式真空过滤机
municipal *adj.* 市政的
dewatering 脱水作用

M10-18
单词读音及例句

M10-19
单词读音及例句

plate-and-frame filter 板框过滤机

zinc *n.* 锌

duct *n.* 导管

M10-20
单词读音及例句

(2) Plate-and-frame filters

Plate-and-frame filters are very commonly used, and consist of porous plates that are held rigidly together in a frame. Plate-and-frame filter presses are perhaps the most widely used type of filtering devices in the chemical industry. The chemical industry uses the filter press in order to separate the solid portion of a chemical slurry from the liquid. A chemical, for example zinc, builds up on the frames. The filter press is then opened and the wet cake, containing solid zinc, can be collected, removed and dried.

Regarding the filter press, feed slurry is pumped to the unit under pressure and flows in the press and into the bottom-corner duct of the frame (Figure 10-7).

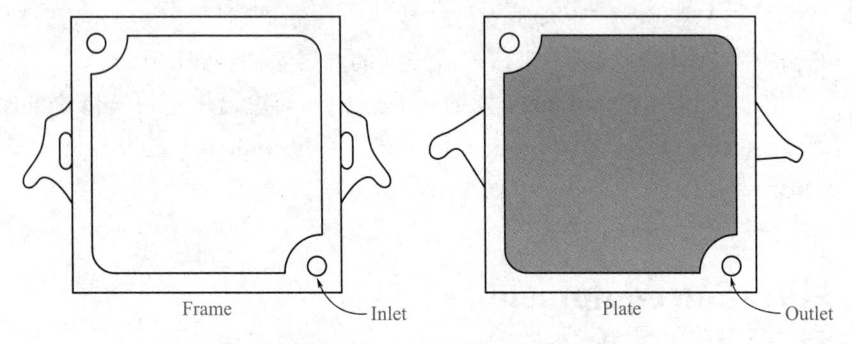

Figure 10-7 Plate-and-frame schematics

hydraulically *adv.* 液压地，水力地

abruptly *adv.* 突然地

press *n.* 压滤机

M10-21
单词读音及例句

This duct has outlets into each of the frames, so the slurry fills the frames in parallel. The plates and frames are assembled alternately with filter cloths over each side of each plate. The assembly is held together as a unit by mechanical force applied hydraulically or by a screw. The liquid filtrate then flows through the filter media while the solids build up in a layer on the inside of the frame side of the media.

The filtrate flows between the filter cloth and the face of the plate to an outlet duct. As filtration proceeds, a cake builds up on the filter cloth until the cake being formed on each face of the frame meet in the center. When this happens, the flow of filtrate, which has been decreasing continuously as the cakes build up, drops off abruptly to a trickle. Filtration is usually stopped well before this occurs.

Notes:

When this happens, the flow of filtrate, which has been decreasing continuously as the cakes build up, drops off abruptly to a trickle.

译文：当这种情况发生时，随着滤饼的堆积而不断减少的滤液流量突然下降为涓流。

语法：句中 when 引导时间状语从句；which 为关系代词，引导非限制性定语从句，修饰 the flow of filtrate。

The process of slurry flow through the press can be seen in Figure 10-8. The slurry enters the lower right-hand side of Frame 18. When Frame 18 is filled with slurry, the excess slurry is forced through the filler media to the upper right hand side of Plate 17. When the slurry is in Plate 17, cake builds up on the filter media. It then flows into Frame 16. When in Frame 16, the slurry flows down because of gravity. When the frame is filled with slurry, the cake forms and the filtrate is sent to the upper right-hand side of Frame 16 and out through Plate 15. Plate 15 leads to the filtrate collecting drum outside of the press.

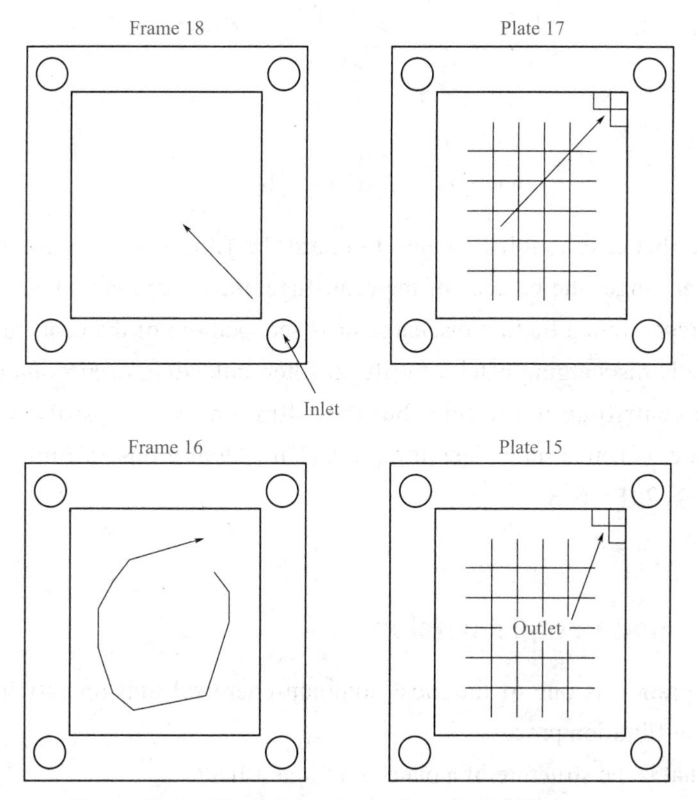

Figure 10-8　Flow of the slurry through the filter press

(3) Centrifugal filters

A filter operation can be carried out using centrifugal force rather than the pressure force used in the equipment described above. Filters using centrifugal force are generally used for coarse granular or crystalline solids and are available primarily for batch operations (Figure 10-9).

Batch centrifugal filters most commonly consist of a basket with perforated sides rotated around a vertical axis. The slurry is fed into the center of the rotating basket and is forced against the basket sides by centrifugal force. There, the liquid passes through the filter medium, which is placed around the inside surface of the basket, and is caught in a "shielding" vessel. The solid phase builds up a filter cake against the filter medium.

coarse　*adj.* 粗糙的

granular　*adj.* 颗粒状的

perforated　*adj.* 穿孔的

vertical axis　纵轴

M10-22
单词读音及例句

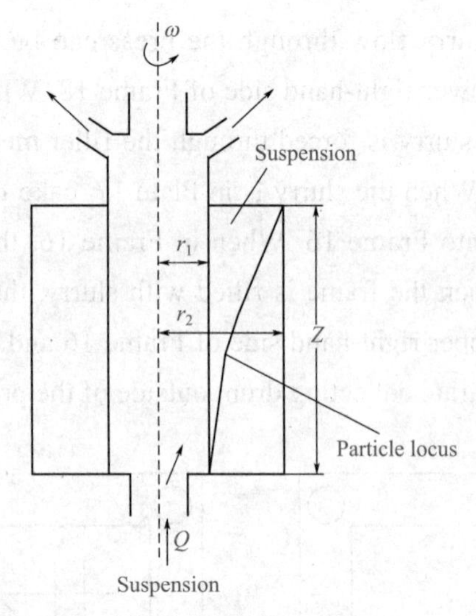

Figure 10-9　Batch centrifugal filter

retard　*v.*
延迟，阻止
halt　*v.*
停止，停顿

When this cake is thick enough to retard the filtration to an uneconomical rate or to endanger the balance of the centrifuge, the operation is halted, and the cake is scraped into a bottom discharge or is scooped out of the centrifuge. In an automatically discharging batch centrifugal filter, unloading occurs automatically while the centrifuge is rotating, but the filtration cycle is still operated in batch mode. (From Louis Theodore, *etc.*, Unit Operations in Environmental Engineering, 2017, 658）

M10-23
单词读音及例句

Reading comprehension

1. Filtration is one of the most common chemical unit operations, please describe the filtration process.

2. What is the structure of a plate and frame filter?

3. What are the advantages of centrifugal filtration?

Exercise

1. True or False.

（1）The really effective filter medium in filter cake filtration are the successive layers added to the cake.

（2）No matter what kind of filter it is, it is operated in batch mode.

（3）When the liquid phase is highly viscous, or when the solids are so fine that vacuum filtration is too slow, pressure filtration can meet the separation requirements.

2. Translate the following sentences into English or Chinese.

（1）在各种过滤设备中，板框压滤机可能是单位过滤面积最便宜的。

____

____

（2）As filtration continues, these particles are thought to bridge across the pores as the cake begins to form on the face of the medium.______________

____

（3）The pharmaceutical industry also uses filter presses in similar applications to concentrate products from process slurries.______________

____

小提示：

• per unit of filtering surface　单位过滤面积

• be thought to　被认为是

Further Reading

Microfiltration

In bioprocesses, a variety of apparatus that incorporate artificial (usually polymeric) membranes are often used for both separations and bioreactions. Microfiltration (MF) is a process that is used to filter very fine particles (smaller than several microns) in a suspension by using a membrane with pores that are smaller than the particles. The driving potential here is the difference in hydraulic pressure.

microfiltration　*n.* 微滤

artificial　*adj.* 人造的

micron　*n.* 微米

M10-24
单词读音及例句

The cells and cell lysates (fragments of disrupted cells) can be separated from the soluble components by using microfiltration with membranes. This separation method offers following advantages:

- It does not depend on any density difference between the cells and the media.
- The closed systems used are free from aerosol formation.
- There is a high retention of cells (>99.9%).
- There is no need for any filter aid.

Depending on the size of cells and debris, and the desired clarity of the filtrate, microfiltration membranes with pore sizes ranging from 0.01 to 10 μm can be used. In cross-flow filtration (CFF; shown in Figure 10-10), the liquid flows parallel to the membrane surface and so provides a higher filtration flux than does dead-end filtration, where the liquid path

lysate　*n.* 溶菌产物

aerosol　*n.* 气溶胶

debris　*n.* 碎片

flux　*n.* 流量

bulk flow　总体流动

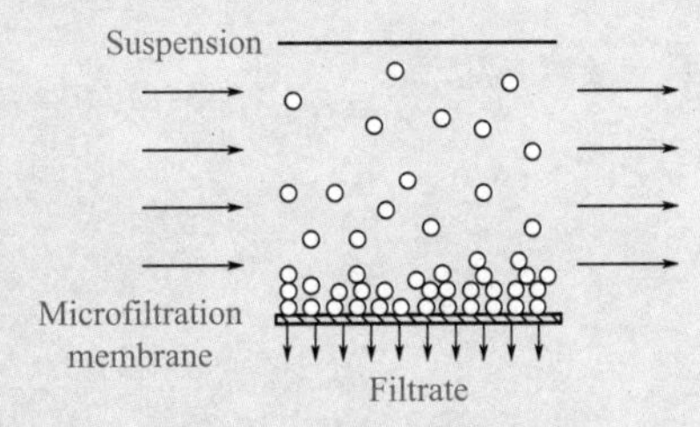

Figure 10-10　Cross-flow filtration

M10-25
单词读音及例句

is solely through the membrane.

In CFF, a lesser amount of the retained species will accumulate on the membrane surface, as some of the retained species is swept from the membrane surface by the liquid flowing parallel to the surface. The thickness of the microparticle layer on the membrane surface depends on the balance between the particles transported by the bulk flow toward the membrane and the sweeping away of the particle by the cross-flow along the membrane.

Reading comprehension

1. What is microfiltration?
2. Please list the advantages of microfiltration.

第十一单元　能量传递及其应用

Unit 11　Energy Transfer and Its Application

Lesson one: Heat Exchanger

In this lesson you will learn:

- Definition of heat exchanger
- Construction of heat exchangers
- Classification of heat exchangers

1. Introduction

A heat exchanger is a heat-transfer device that is used for the transfer of internal thermal energy between two or more fluids available at different temperatures. In most heat exchangers, the fluids are separated by a heat-transfer surface and ideally they do not mix. Heat exchangers are used in the petroleum, transportation, air conditioning, refrigeration, heat recovery, and other industries. Common examples of heat exchangers familiar to us in day-to-day use are automobile radiators, condensers, evaporators air preheaters and oil coolers. Heat exchangers could be classified in many different ways.

exchanger *n.* 交换器

thermal *adj.* 热的，保温的

heat-transfer surface 传热面

radiator *n.* 散热器

condenser *n.* 冷凝器

evaporator *n.* 蒸发器

M11-1
单词读音及例句

Notes:

A heat exchanger is a heat-transfer device that is used for the transfer of internal thermal energy between two or more fluids available at different temperatures.

译文：换热器是一种传热装置，用于在不同温度下两种或多种液体之间传递内部热能。

语法：句中 that 后面引导定语从句，用来说明 device 的具体内容，是英语语法中经常用到的关系代词。

2. Construction of heat exchangers

A heat exchanger consists of heat-exchanging elements such as a core or matrix containing the heat-transfer surface, and fluid distribution elements such as headers or tanks, inlet and outlet nozzles or pipes, *etc*. Usually, there are no moving parts in the heat exchanger, however there are exceptions, such as a rotary regenerator in which the matrix is driven to rotate at some design speed. The heat-

matrix *n.* 基体，矩阵
header *n.* 集流管
nozzle *n.* 管嘴，喷嘴
conduction *n.* 传导
fin *n.* 尾翅，散热片

M11-2
单词读音及例句

transfer surface is in direct contact with fluids through which heat is transferred by conduction. The portion of the surface that separates the fluids is referred to as the primary or direct contact surface. To increase heat-transfer area, secondary surfaces known as fins may be attached to the primary surface.

Notes:

A heat exchanger consists of heat-exchanging elements such as a core or matrix containing the heat-transfer surface, and fluid distribution elements such as headers or tanks, inlet and outlet nozzles or pipes, *etc*.

译文：换热器由换热元件，如包含传热面的芯或管网和流体输送元件，如头或罐、进口和出口喷嘴或管道等组成。

语法：containing 在这里是现在分词，是“包含”的意思，补充修饰前面的 core 和 matrix。

3. Classification of heat exchangers

tubular heat exchangers 管式换热器
plate heat exchangers 板式换热器
coiled *adj.* 盘绕的
gasketed *adj.* 封圈型的
lamella *n.* 薄板，薄片
regenerator *n.* 蓄热式回热器
concentric *adj.* 同轴的
inventory *n.* 存货
longitudinal *adj.* 纵向的
circumferential *adj.* 圆周的

According to constructional details, heat exchangers are classified as:

- Tubular heat exchangers-double pipe, shell and tube, coiled tube;
- Plate heat exchangers- gasketed, spiral, plate coil, lamella;
- Extended surface heat exchangers;
- Regenerators.

(1) Double pipe exchangers

A double pipe heat exchanger has two concentric pipes, usually in the form of a U-bend design as shown in Figure 11-1. The flow arrangement is pure countercurrent. A number of double pipe heat exchangers can be connected in series or parallel as necessary. Their usual application is for small duties requiring, typically, less than 300 ft^2 and they are suitable for high pressures and temperatures, and thermally long duties. This has the advantage of flexibility since units can be added or removed, as required, and the design is easy to service and requires low inventory of spares because of its standardization. Either longitudinal fins or circumferential fins within the annulus on the inner pipe wall are required to enhance the heat transfer from the inner pipe fluid to the annulus fluid.

M11-3
单词读音及例句

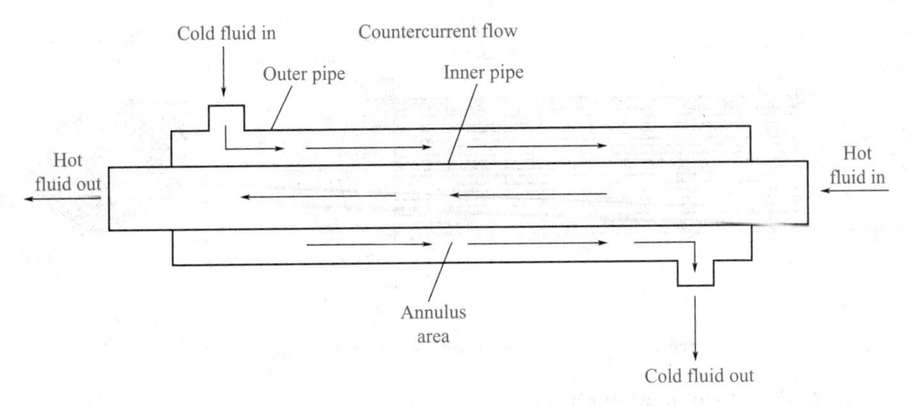

Figure 11-1　Double pipe heat exchanger schematic

(2) Shell and tube heat exchangers

In process industries, shell and tube heat exchangers are used in great numbers, far more than any other type of exchanger. More than 90% of heat exchangers used in industry are of the shell and tube type. The shell and tube heat exchangers are the "work horses" of industrial process heat transfer. They are the first choice because of well-established procedures for design and manufacture from a wide variety of materials, many years of satisfactory service, and availability of codes and standards for design and fabrication. They are produced in the widest variety of sizes and styles. There is virtually no limit on the operating temperature and pressure.

shell and tube heat exchangers 管壳式换热器

fabrication　*n.* 制造

M11-4
单词读音及例句

Notes:

In process industries, shell and tube heat exchangers are used in great numbers, far more than any other type of exchanger.

译文：在流程工业中，管壳式换热器被大量使用，远远超过任何其他类型的换热器。

语法：这里的重点语法是强化式比较结构，"far more than..."，"远远超过"的意思。

The simplest shell and tube heat exchanger has a single pass through the shell and a single pass through the tubes (Figure 11-2). Fluids that flow through tubes at low velocity result in low heat transfer coefficients and low-pressure drops. To increase the heat transfer rates, multi-pass operations may be used. Baffles are used to divert the fluid within the distribution header. An exchanger with one pass on the shell side and four tube passes is termed a 1-4 shell and tube heat exchanger. It is also possible to increase the number of passes on the shell side by using dividers.

multi-pass　*n.* 多程

baffle　*n.* 挡板

M11-5
单词读音及例句

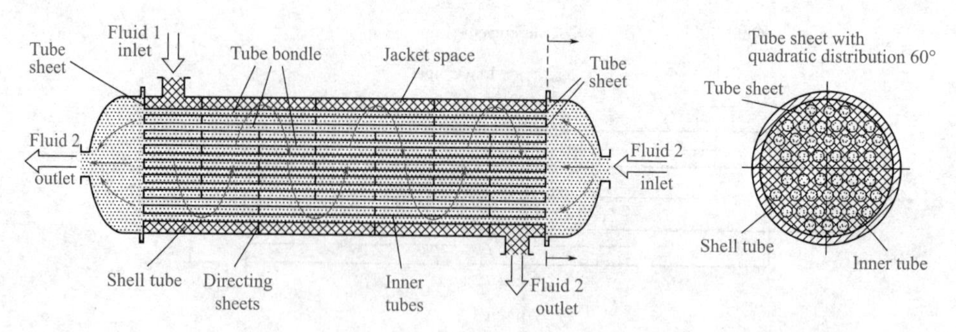

Figure 11-2 Shell and tube heat exchanger schematic

small-bore *adj.* 小口径的

ductile *adj.* 柔软的

helix *n.* 螺旋状物

M11-6 单词读音及例句

(3) Coiled tube heat exchanger

One of the three classical heat exchangers used today for large-scale liquefaction systems is the coiled tube heat exchanger(CTHE) (see Figure 11-3). The construction of these heat exchangers involves winding a large number of small-bore ductile tubes in helix fashion around a central core tube, with each exchanger containing many layers of tubes along both the principal and radial axes. Tubes in individual layers or groups of layers may be brought together into one or more tube plates through which different fluids may be passed in counter-flow to the single shellside fluid.

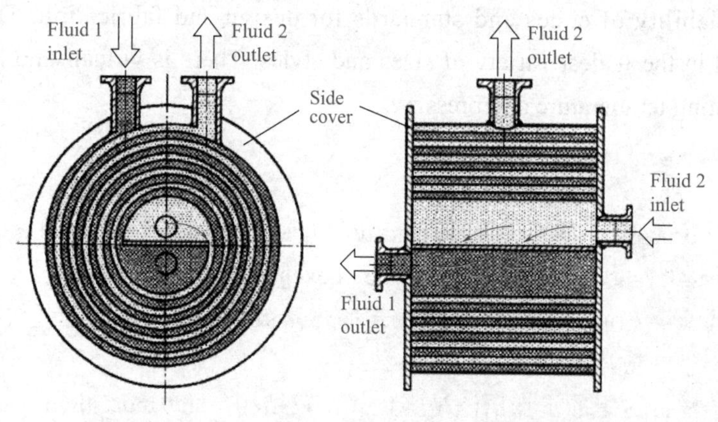

Figure 11-3 Coiled tube heat exchanger schematic

shellside *n.* 壳程

aluminum *n.* 铝

alloy *n.* 合金

cryogenic *adj.* 冷冻的，低温的

M11-7 单词读音及例句

The high-pressure stream flows through the small-diameter tubes, while the low-pressure return stream flows across the outside of the small-diameter tubes in the annular space between the inner central core tube and the outer shell. Pressure drops in the coiled tubes are equalized for each high-pressure stream by using tubes of equal length and varying the spacing of these in the different layers. Because of small-bore tubes on both sides, CTHEs do not permit mechanical cleaning and therefore are used to handle clean, solid-free fluids or fluids whose fouling deposits can be cleaned by chemicals. Materials are usually aluminum alloys for cryogenics, and stainless steels for high-temperature applications.

The coiled tube heat exchanger offers unique advantages, especially when dealing with low-temperature applications for the following cases:

- Simultaneous heat transfer between more than two streams is desired;
- A large number of heat-transfer units is required;
- High operating pressures are involved.

The coiled tube heat exchanger is not cheap because of the material costs, high labor input in winding the tubes, and the central mandrel, which is not useful for heat transfer but increases the shell diameter.

Reading comprehension

1. What is a heat exchanger used for?

2. How to increase heat-transfer area?

3. How many classifications of heat exchangers are there according to constructional details?

4. Why are shell and tube heat exchangers the first choice for process industries?

5. Why is the coiled tube heat exchanger costly?

Exercise

1. Match the items listed in the following two columns.

Heat exchanging element	表面紧凑程度
Fluids distribution element	流体输送元件
Degree of surface compactness	环形空间
The annular space	换热元件
A form of kinetic energy	非接触材料
Non-conducting substance	动能的一种形式

2. Translate the following sentences into English or Chinese.

（1）The conductivities of materials vary widely, being greatest for metals, less for nonmetals, still less for liquids, and least for gases.________________

____

（2）Transfer of energy by radiation is unique in that non-conducting substance is necessary, as with conduction and convection.________________

____

（3）流体有相变时的对流传热系数远远大于无相变时的对流传热系数。

____

____

（4）高温液体走红色管道，低温液体走蓝色和绿色管道的中间部位。

____

____

小提示：

• 强化比较句式：Far more than...

• 相变：phase change 或者 phase transition

Further Reading

Heat Transfer

convection *n.* 对流

radiation *n.* 热辐射

soldering iron 焊铁

opaque *adj.* 不透明的

quartz *n.* 石英

insulator *n.* 绝缘体

M11-8
单词读音及例句

Heat, a form of kinetic energy, is transferred in three ways: (i) Conduction; (ii) Convection; (iii) Radiation. Heat can be transferred only if a temperature difference exists, and then only in the direction of decreasing temperature. Beyond this, the mechanisms and laws governing each of these ways are quite different.

Conduction

Heat flows by conduction from the soldering iron to the work, through the brick wall of a furnace, through the wall of a house, or through the wall of a cooking utensil. Conduction is the only mechanism for the transfer of heat through an opaque solid.

Some heat may be transferred through transparent solids, such as glass, quartz, and certain plastics, by radiation. In fluids, the conduction is supplemented by convection, and if the fluid is transparent, by radiation. The conductivities of materials vary widely, being greatest for metals, less for nonmetals, still less for liquids, and least for gases. Any material that has a low conductivity may be considered to be an insulator. This increases the length of path for heat flow and at the same time reduces the effective cross-sectional area, both of which decrease the heat flow.

Convection

Heat convection involves the transfer of heat by the mixing of molecules of a fluid with the body of the fluid after they have either gained or lost heat by intimate contact with a hot or cold surface. The transfer of heat at the hot or cold surface is by conduction. For this reason, heat transfer by convection cannot occur without conduction.

The motion of the fluid to bring about mixing may be entirely due to differences in density resulting from temperature differences, as in natural convection, or it may be brought about by mechanical means, as in forced convection.

A steam or hot-water radiator is a good example of heating by convection. Air in contact with the radiator warms and rises because it is lighter than the other air in the room. It is displaced by cool air. In fact, the heat from the fire in the furnace heating the hot water or steam is transferred to the boiler wall by convection, and the hot water or steam transfers heat from the boiler to the radiator by convection.

Radiation

Solid material, regardless of temperature, emits radiations in all directions. These radiations may be, to varying degrees, absorbed, reflected, or transmitted. The lost energy that is transferred by radiation is equal to the difference between the radiations emitted and those absorbed.

The radiations from solids form a continuous spectrum of considerable width, increasing in intensity from a minimum at a short wavelength through a maximum and then decreasing to a minimum at a long wavelength. As the temperature of the object is increased, the entire emitted spectrum decreases in wavelength. As the temperature of an iron bar, for example, is raised to about 500 ℃, the radiations become visible as a dark red glow. As the temperature is increased further, the intensity of the radiation increases and the color becomes more yellow.

Liquids and gases only partially absorb or emit these radiations, and do so in a selective fashion. Many liquids, especially organic liquids, have selective absorption bands in the infrared and ultraviolet regions.

net energy　净能量

radiator　*n.*
散热器，暖气片

emit　*vt.*
放射，发出，散发

transmit　*vt.*
传导，转送

spectrum　*n.*
光谱，频谱

infrared　*adj.*
红外线的

ultraviolet　*adj.*
紫外线的

M11-9
单词读音及例句

Reading comprehension

1. How many ways can heat be transferred and what are they?
2. What is the only mechanism for the transfer of heat through an opaque solid?
3. What is the net energy that is transferred by radiation?

第十二单元　质量传递及其应用

Unit 12　Mass Transfer and Its Application

Lesson one: Absorption

In this lesson you will learn:

- Define absorption and stripping
- Packed columns
- Packing

1. Absorption and stripping

The removal of one or more selected components from a gas mixture by absorption is one of the most important operations in the field of mass transfer.

(1) Absorption

Absorption is defined as the process of transfer of molecules of one substance directly to another substance. It may be either a physical or a chemical process. Physical absorption depends on the solubility of the substance absorbed and chemical absorption involves chemical reactions between the absorbed substance and the absorbing medium. The process of absorption conventionally refers to the intimate contacting of a mixture of gases with a liquid so that part of one or more of the constituents of the gas will dissolve in the liquid. It is a gas-liquid transfer process. It is used in the selective removal of a gaseous contaminant or product from a gas mixture. Removal is often effected by absorption in a liquid in which only the gas concerned is soluble.

absorption *n.* 吸收

stripping *n.* 气液分离

constituent *n.* 成分

contaminant *n.* 杂质，污染物

soluble *adj.* 可溶的

M12-1
单词读音及例句

Notes:

The process of absorption conventionally refers to the intimate contacting of a mixture of gases with a liquid so that part of one or more of the constituents of the gas will dissolve in the liquid.

译文：吸收过程通常是指气体混合物与液体的密切接触，使得气体的一种或多种成分的一部分溶解在液体中。

语法：refer to 意为“指的是”，常用于描述定义的句子中；so that 引导状语从句，意为“以便，以使”。

The contact usually occurs in some type of packed column, although plate and spray towers are also used. In gas absorption operations the equilibrium of absorption is that between a relatively nonvolatile absorbing liquid (solvent) and a soluble gas (solute) (Figure 12-1). The usual operating data to be determined or estimated for isothermal packed tower systems are the liquid rate(s) and the terminal concentrations or mole fractions. An operating line, that describes operating conditions in the column, is obtained by a mass balance around the column.

packed column 填料塔

spray tower 喷雾塔

nonvolatile *adj.* 非挥发性的

isothermal *adj.* 等温的

M12-2 单词读音及例句

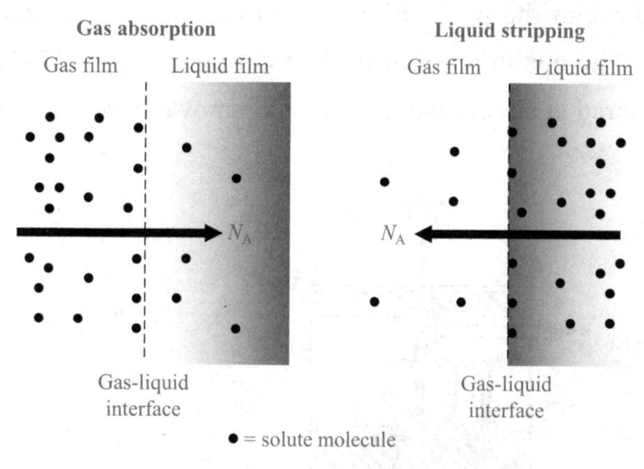

Figure 12-1　Gas absorption *vs.* liquid stripping for transfer of solute A between contacting gas and liquid phases

(2) Stripping

It is sometimes also necessary to free a liquid from dissolved gases. Quite often, an absorption column is followed by a liquid absorption process in which the gas solute is removed from the absorbing medium by contact with an insoluble gas. This reverse operation is called stripping and is utilized to generate the rich solvent so that it may (in many cases) be recycled back to the absorption unit. The rich solution enters the stripping unit and the volatile solute is stripped from solution by either reducing the pressure, increasing the temperature, using a stripping gas to remove the vapor solute dissolved in the solvent, or any combination of these process changes. While the concept of stripping is opposite to that of absorption, it is treated in the same manner.

solute *n.* 溶质，溶解物

stripping unit 汽提装置

M12-3 单词读音及例句

Notes:

This reverse operation is called stripping and is utilized to generate the rich solvent so that it may (in many cases) be recycled back to the absorption unit.

译文：这种反向操作称为汽提，用于生成富溶剂，以便（在许多情况下）可以循环回吸收装置。

语法：句中并列连词 and 连接两个被动语态结构 is called... 和 is utilized...; so that 引导状语从句，意为“以便，以使”。

2. Packed columns

trickle v. 滴

expose v. 暴露，遭遇

M12-4
单词读音及例句

The principal types of gas absorption equipment are packed columns (continuous operation) and plate columns (stage operation). Of the two categories, the packed column is most commonly used. Plate columns will also receive treatment in the next lesson (Distillation).

Packed columns are usually vertical columns that have been filled with packing or material of large surface area. The liquid is distributed over and trickles down through the packed bed, thus exposing a large surface area to contact the gas. The countercurrent packed column (Figure 12-2) is the most common encountered in environmental gaseous removal or recovery systems.

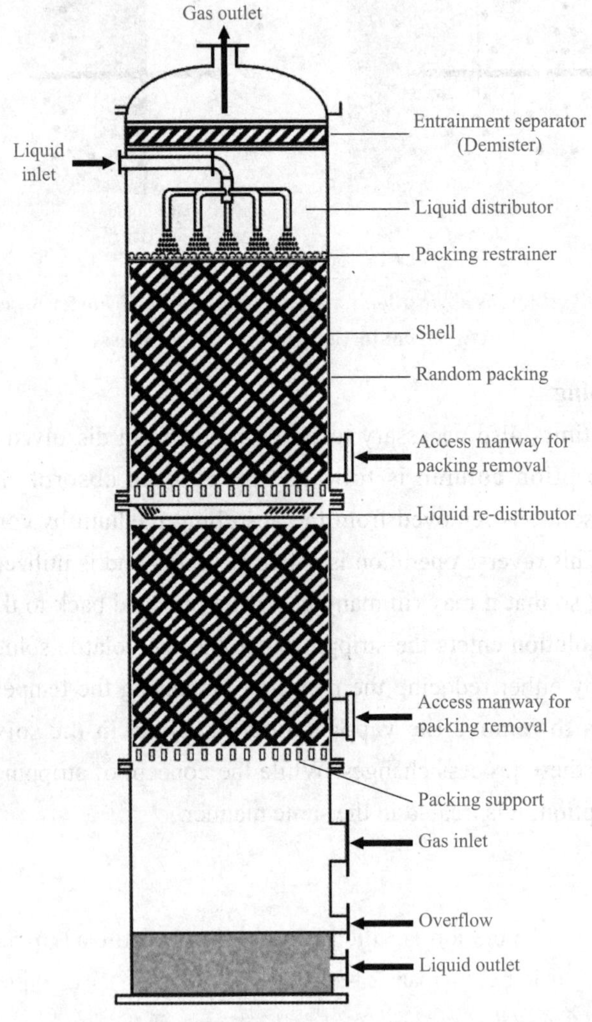

Figure 12-2 Typical countercurrent packed column

The gas stream moves upward through the packed bed against an absorbing or reacting liquor (solvent-scrubbing solution), which is introduced at the top of the packing. This results in the highest possible efficiency. Since the solute concentration in the gas stream decreases as it rises through the column, there is fresh solvent constantly available for contact. This provides the maximum average driving force for the mass transfer process throughout the packed bed. Mist eliminators also play an important role in absorbers. Mist eliminators are used to remove liquid droplets entrained in thc gas stream. Ease of separation depends on the size of the droplets. Droplets formed from liquids are usually large, up to hundreds of microns in diameter, and are therefore effectively removed in mist eliminators. However, drops formed due to condensation or chemical reactions may be less than 1 μm in size and much harder to separate.

liquor　*n.* 溶液

scrub　*v.* 洗涤，擦洗

mist eliminator 除雾器

entrain　*v.* 带走，夹带

M12-5
单词读音及例句

Notes:

The gas stream moves upward through the packed bed against an absorbing or reacting liquor (solvent-scrubbing solution), which is introduced at the top of the packing.

译文：气流通过填料床层向上移动，与从填料顶部引入的吸收或反应液（溶剂洗涤溶液）反向。

语法：句中 which 为关系代词，引导非限制性定语从句，修饰 an absorbing or reacting liquor。

Packed columns are characterized by a number of features to which their widespread popularity may be attributed.

- Minimum structure—the packed column usually needs only a packing support and liquid distributor approximately every 10 ft along its height.
- Versatility—the packing material can be changed by simply discarding it and replacing it with a type providing better efficiency.
- Corrosive-fluids handling—ceramic packing is used and may be preferable to metal or plastic because of its corrosion resistance; when packing does deteriorate, it can be quickly and easily replaced, and it is also preferred when handling hot combustion gases.
- Low pressure drop—unless operated at very high liquid rates where the liquid becomes the continuous phase as the flowing films thicken and merge, the pressure drop per lineal foot of packed height is relatively low.
- Range of operation—although efficiency varies with gas and liquid feed rates, the range of operation is relatively broad.
- Low investment—when plastic packings are satisfactory or when the columns are less than about 3 or 4 feet in diameter, cost is relatively low.

versatility　*n.* 多功能性

corrosive　*adj.* 腐蚀性的

ceramic　*adj.* 陶瓷的

deteriorate　*v.* 变坏，退化

entail　*v.* 需要

operational　*adj.* 操作的，运作的

M12-6
单词读音及例句

3. Packing

The packing is the heart of this type of equipment. Its proper selection entails an understanding of the operational characteristics and the effect on performance of the significant physical differences among the various packing types (Figure 12-3). The main points to be considered in choosing the column packing include:

- Durability and corrosion resistance.
- Free space per unit volume of packed space.
- Wetted surface area per unit volume of packed space.
- Resistance to the flow of gas.
- Packing stability and structural strength to permit easy handling and installation.
- Weight per unit volume of packed space.
- Cost per unit area of packed space.

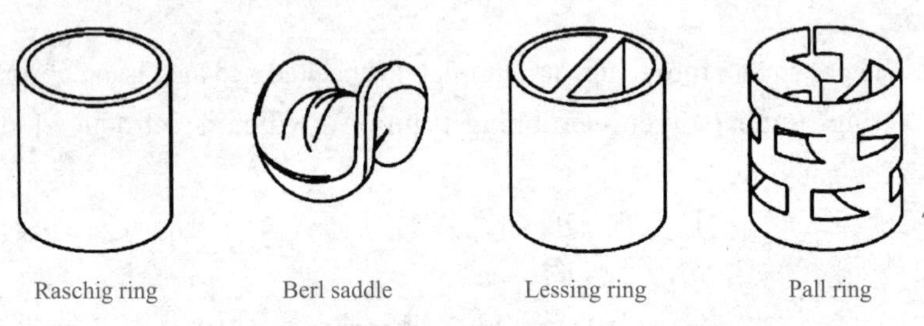

Figure 12-3 Common industrial tower packing

Raschig ring
拉西环
Berl saddle
贝尔鞍
side thrust 侧压
flooding point
泛点
Pall ring 鲍尔环

M12-7
单词读音及例句

Raschig ring

Originally, the most popular type, usually cheaper per unit cost but sometimes less efficient than others.

Berl saddle

More efficient than Raschig ring in most applications, but more costly; do not produce much side thrust and have lower unit pressure drops with higher flooding points than Raschig ring; easier to break in than Raschig ring.

Lessing ring

One of the most efficient packings, but more costly; very little tendency or ability to nest and block areas of bed; higher flooding limits and lower pressure drop than Raschig ring or Berl saddle; easier to break in than Raschig ring.

Pall ring

Lower pressure drop, less than half of Raschig ring; higher flooding limit; good liquid distribution; high capacity; considerable side thrust on column wall.

random packing
散装填料
stacked packing
规整填料
tedious *adj.*
沉闷的

M12-8
单词读音及例句

One additional distinction should be made, *i.e.*, the difference between random and stacked (structured) packings. Random packings are those that are simply dumped into the column during installation and allowed to fall at random. It is the most common method of packing installation. Stacked packing, on the other hand, is specifically laid out and stacked by hand, making it a tedious operation and rather costly; it is avoided where possible except for the initial layers on supports. A good packing from a process viewpoint can be reduced in effectiveness by poor liquid distribution across the top of its upper surface.

Notes:

Random packings are those that are simply dumped into the column during installation and allowed to fall at random.

译文：散装填料是那些在安装过程中简单地倾倒在塔中并允许随机落下的填料。

语法：句中 that 为关系代词，引导定语从句，修饰 those，而指示代词 those 又指代 packings。

Reading comprehension

1. What is absorption?
2. How to obtain the operating line of the packed tower?
3. How to get rich solvent by stripping?
4. What are packed columns?
5. What is the importance of the packing in the packed columns?

Exercise

1. True or False.

(1) In gas absorption operations the equilibrium of absorption is that between a relatively nonvolatile absorbing liquid and a soluble gas.

(2) Droplets formed due to condensation or chemical reactions may be less than 1 μm in size and are effectively removed in mist eliminators.

(3) Stacked packing is avoided where possible except for the initial layers on supports.

2. Translate the following sentences into English or Chinese.

(1) 在气体吸收操作中，特殊溶剂的选择也很重要。________________

____

(2) 安装过程中，在将填料倒入柱体之前柱内应装入水。__________

____

(3) The redistribution brings the liquid off the wall and directs it toward the center of the column for redistribution and contact in the next lower section.

____

____

小提示：
prior to... 在……之前
be filled with 用……装满

Further Reading

Adsorption

adsorption *n.* 吸附

adsorbent *n.* 吸附剂

zeolite *n.* 沸石

granular *adj.* 颗粒的

M12-9
单词读音及例句

Adsorption is a mass transfer process in which a solute is removed from a gas stream because it adheres to the surface of a solid. In an adsorption system (Figure 12-4), the gas stream is passed through a layer of solid particles referred to as the adsorbent bed. As the gas stream passes through the adsorbent bed, the solute absorbs or "sticks" to the surface of the solid adsorbent particles. Eventually, the adsorbent bed becomes "filled" or saturated with the solute. The adsorbent, for example, activated carbon, molecular sieves, zeolite, clay, or silica gel, is generally used in granular form in a fixed bed and less often used as a suspension in a liquid.

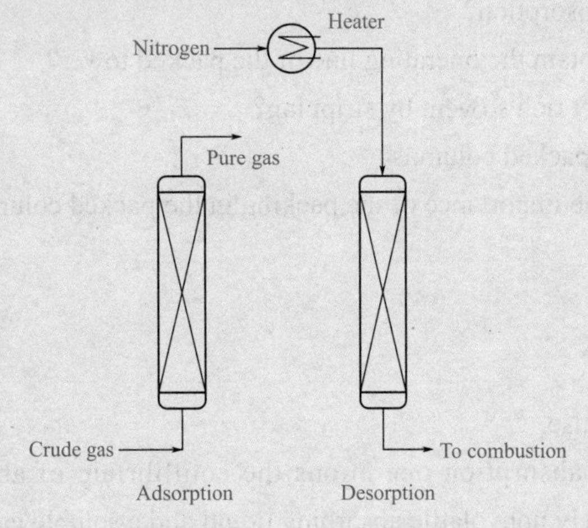

Figure 12-4 Adsorber setup

chemisorption *n.* 化学吸附

separability *n.* 分离性

M12-10
单词读音及例句

Desorption is usually carried out by physical methods such as temperature increase, pressure drop, or exchange against another medium. In the case of high binding energies (>50 kJ/mol, chemisorption), physical desorption is no longer possible. For suspended adsorbents, stirred vessels are generally used, and the process is easily designed. For scale-up, the erosion behavior of the adsorbent and its separability from the liquid are important. The amount of energy to be introduced is determined experimentally and should be high enough that suspension just takes place. Slow-running stirrers with propeller or inclined blades of large diameter are favorable.

In most cases adsorbents are used in the form of fixed beds. A number of theoretical treatments are available for the adsorption equilibrium and the adsorption kinetics. However, it is recommended to determine these parameters, the possible loading, and the breakthrough behavior experimentally. This also applies to the desorption process.

In continuous processes at least two adsorbers are used in parallel (adsorption, desorption). Because of the complicated operation, fully automatic apparatus is preferred. (From G. Herbert Vogel, Process Development, 2005, 477）

Reading comprehension

1. What is the difference between adsorption and absorption?
2. How to perform the desorption?

Lesson two: Distillation

In this lesson you will learn:

- Definition of distillation
- Batch distillation
- Continuous distillation

1. Introduction

Distillation is no doubt the most widely used separation process in the chemical industries. Applications range from the rectification of alcohol to the fractionation of crude oil. Distillation may be defined as the separation of the components of a liquid feed mixture by a process involving partial vaporization through the application of heat. In general, the vapor evolved is recovered in liquid form by condensation. The degree of difference in relative volatility of the components dictates the extent of separation of the mixture. The more volatile (lighter) components of the liquid mixture are obtained in the vapor discharge at a higher concentration.

Distillation separates components with different volatilities, that is with different boiling points. Pure water, at atmospheric pressure, boils at 100 ℃; at this temperature, water and its vapor are in equilibrium. Assume that we have to separate the water-methanol mixture. Methanol boils at 64.7 ℃ at atmospheric pressure. Figure 12-5 shows the isobaric vapor-liquid equilibrium data of the water-methanol mixture.

distillation *n.* 蒸馏

rectification *n.* 精馏

fractionation *n.* 分馏

vaporization *n.* 汽化

volatility *n.* 挥发性

methanol *n.* 甲醇

isobaric *adj.* 等压的

liquidus *n.* 液相线

M12-11
单词读音及例句

M12-12
单词读音及例句

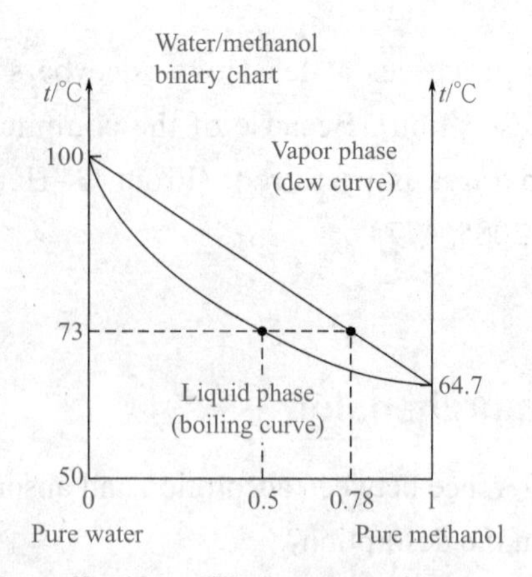

Figure 12-5　Vapor-liquid equilibrium data of water-methanol mixture

A mixture containing 50% (mol) of methanol will boil at 73℃. The vapor in equilibrium with the liquid phase (liquidus) will have a molar composition of 78% in methanol. It is therefore richer in methanol than the liquid. This property is used in distillation by putting liquid and vapor into contact by means of counter-current.

2. Batch distillation

batch distillation 间歇蒸馏

counterpart *n.* 对应的人或物

constraint *n.* 限制

M12-13
单词读音及例句

Although batch distillations are generally more costly than their continuous counterparts, there are certain applications in which batch distillation is the method of choice. Batch distillation is typically chosen when it is not possible to run a continuous process due to limiting process constraints, the need to distill other process streams, or because the low frequency use of distillation does not warrant a unit devoted solely to a specific product or operation. It is highly flexible, since it can easily be combined with other process steps.

Notes:

Batch distillation is typically chosen when it is not possible to run a continuous process due to limiting process constraints, the need to distill other process streams, or because the low frequency use of distillation does not warrant a unit devoted solely to a specific product or operation.

译文：当由于工艺限制、蒸馏其他工艺流的需要，或由于蒸馏的低频使用不保证一个装置仅用于特定产品或操作，无法运行连续工艺时，通常选择分批蒸馏。

语法：句中 when 为连词，引导时间状语从句；it is not possible to... 结构中，it 为先行主语，代表后移的不定式 to run...；句中第二个逗号后的并列连词 or 连接原因状语 due to limiting... other process streams 和原因状语从句 because the low frequency use... operation。

Batch distillation is generally carried out in the upward operating mode, in which the initial mixture is heated in the distillation vessel and the individual components are collected overhead one after the other in order of their volatilities, starting with the lowest boiling fraction (Figure 12-6).

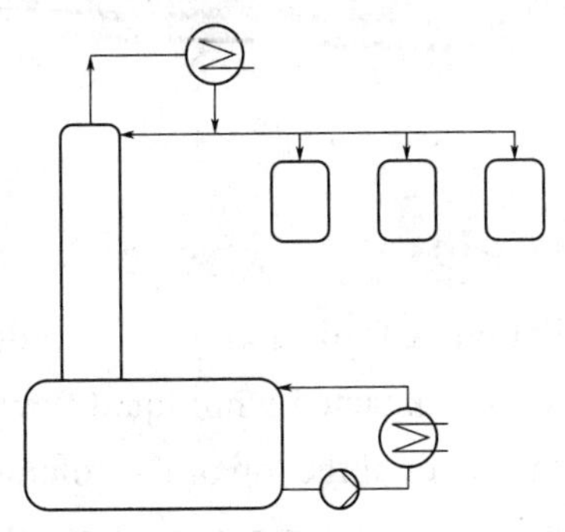

Figure 12-6 Batch distillation with upward operating mode

The reflux ratio in a batch distillation should not be kept constant during the separation of a fraction, but should gradually be increased to ensure constant purity of the overhead product with time. At a constant reflux ratio the purity of the overhead product gradually decreases with time, since the content of the component being separated in the distillation vessel decreases and the separation thus becomes more difficult. The resulting savings are considerable. In comparison to operation with constant reflux ratio, a reduction in the distillation time and energy requirement of about 30%-60% can be expected.

reflux ratio 回流比

overhead product 塔顶产物

intermediate *adj.* 中间的

M12-14
单词读音及例句

When high purities of the individual fractions are required, it is necessary to collect intermediate fractions, which are stored separately and added to the next batch. The individual intermediate fractions are also best collected with increasing reflux ratio. To reduce the number of intermediate fractions, the separating internals should have as low a liquid holdup as possible.

In comparison to continuous distillation, batch distillation have the disadvantage of higher thermal loading of the product due to the longer residence time, and the energy requirement is also higher.

In some situations, the degree of separation achieved in a single equilibrium stage (such as a flash distillation column) or in a batch still is often not large enough to obtain the desired distillate and/or bottoms purities. To improve the recovery of the desired product, a multi-staged distillation column may be employed. (From G. Herbert Vogel, Process Development, 2005, 477）

liquid holdup 持液量

residence time 停留时间

multi-staged distillation 多级蒸馏

M12-15
单词读音及例句

Notes:

In some situations, the degree of separation achieved in a single equilibrium stage (such as a flash distillation column) or in a batch still is often not large enough to obtain the desired distillate and/or bottoms purities.

译文：在某些情况下，在单个平衡阶段（例如闪蒸塔）或间歇蒸馏器中实现的分离程度通常不足以获得所需的馏出物和（或）底部产物纯度。

语法：句中 achieved… batch still 为过去分词短语作后置定语，修饰 the degree of separation。

3. Continuous distillation

In continuous distillation, a feed mixture is introduced to a column where vapor rising up the column is contacted with liquid flowing downward (which is provided by condensing the vapor at the top of the column). This process removes or absorbs the less volatile (heavier) components from the vapor, thus effectively enriching the vapor with the more volatile (lighter) components.

Notes:

In continuous distillation, a feed mixture is introduced to a column where vapor rising up the column is contacted with liquid flowing downward (which is provided by condensing the vapor at the top of the column).

译文：在连续蒸馏中，将进料混合物引入塔中，在塔中上升的蒸气与向下流动的液体接触（通过冷凝塔顶部的蒸气提供）。

语法：句中 where 为关系副词，引导定语从句，修饰 column；括号中关系代词 which 引导定语从句，修饰 liquid flowing downward。

condenser *n.* 冷凝器

descend *v.* 降落

remainder *n.* 剩余物

columnwise *adj.* 列式的

stripping section 提馏段

rectifying section 精馏段

M12-16
单词读音及例句

In a distillation column, vapor flows up the column and liquid flows countercurrently down the column. The vapor and liquid are brought into contact on plates as shown in Figure 12-7. The vapor from the top of column is sent to a condenser. Part of the condensate from the condenser is returned to the top of the column as reflux to descend counter to the rising vapors. The remainder of the condensed liquid is drawn as product. The ratio of the amount of reflux returned to the column to the distillate product collected is known as the reflux ratio. As the liquid stream descends the column it is progressively enriched with low-boiling constituents. The column internals are then used as an apparatus for bringing these streams into intimate contact so that the vapor stream tends to vaporize the low-boiling constituent from the liquid and the liquid stream tends to condense the high-boiling constituent from the vapor. Columnwise distillation involves two sections: an enriching (top) and stripping (bottom) section, which lies below the feed where the more volatile components are stripped from the liquid and the enriching or rectifying section above the feed.

Notes:

The ratio of the amount of reflux returned to the column to the distillate product collected is known as the reflux ratio.

译文：返回塔的回流量与收集的馏出物产品的比率称为回流比。

语法：句中 returned to the column 和 collected 均为过去分词作后置定语；句中第二个 to 连接比例的前后两部分。

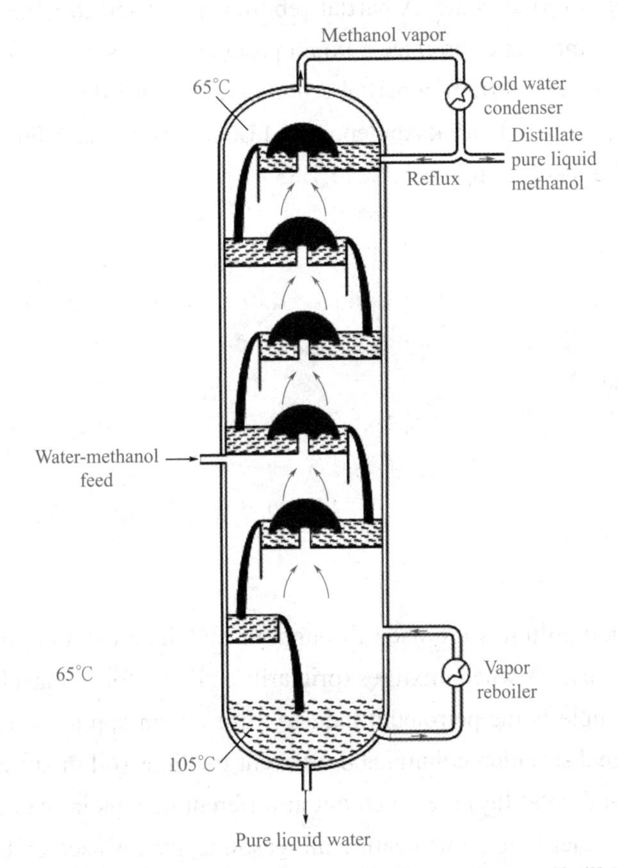

Figure 12-7　Water-methanol separation by continuous distillation

A column may consist of one or more feeds and may produce two or more product streams. The product recovered at the top of a column is referred to as the tops, while the product at the bottom of the column is referred to as the bottoms. Any product (s) drawn at various stages between the top and bottom are referred to as side streams. Multiple feeds and product streams do not alter the basic operation of a column, but they do complicate the calculations and analysis of the process to some extent. If the process requirement is to strip a volatile component from a relatively nonvolatile solvent, the rectifying (bottom) section may be omitted, and the unit is then called a stripping column. Virtually pure top and bottom products can be achieved by using many stages or (occasionally) additional columns; however, this is not usually economically feasible.

omit　*v.* 省略

virtually　*adv.* 事实上

M12-17
单词读音及例句

In some operations where the top product is required as a vapor, the liquid condensed is sufficient only to provide reflux to the column, and the condenser is referred to as a partial condenser. In a partial condenser, the reflux will be in equilibrium with the vapor leaving the condenser, and is considered to be an equilibrium stage in the development of the operating line when estimating the column height. When the liquid is totally condensed, the liquid returned to the column will have the same compositions as the top product and is not considered to be in an equilibrium state. A partial reboiler is utilized to generate vapor to operate the column and to produce a liquid product if necessary. Since both liquid and vapor are in equilibrium, a partial reboiler is considered to be an equilibrium stage as well. (From J. Patrick Abulencia and Louis Theodore, A Future Chemical Engineering Education Approach, 2015, 583）

Notes:

When the liquid is totally condensed, the liquid returned to the column will have the same compositions as the top product and is not considered to be in an equilibrium state.

译文：当液体完全冷凝时，返回塔的液体将具有与塔顶产物相同的组成，并且不被认为处于平衡状态。

语法：句首 when 为连词，引导时间状语从句；the same… as… 意为“与……相同的……”。

jet fuel 航天燃料

gasoline *n.* 汽油

M12-18
单词读音及例句

Distillation columns are used throughout the chemical and environmental engineering industries when mixtures (primarily in liquid form) must be separated. One such example is the petroleum industry. In such an application, crude oil is fed into a large distillation column and different fractions (oil mixtures of varying composition and volatility) are taken out at different heights in the column. Each fraction, such as jet fuel, home heating oil, gasoline, *etc*., is used by both industry and the consumer in a variety of ways. The separation achieved in a distillation column depends primarily on the relative volatilities of the components to be separated, the number of contacting trays (plates) or packing height, and the ratio of liquid and vapor flow rates.

Reading comprehension

1. What will affect the separation of the mixture by distillation?
2. What applications are batch distillation suitable for?
3. Should the reflux ratio in a batch distillation be kept constant during the separation of a fraction?

4. What is the difference between a stripping column and a distillation column?

5. Why should the partial condenser at the top of the distillation column be regarded as an equilibrium stage?

Exercise

1. Noun explanation.

（1）Distillation

（2）Batch distillation

（3）Continuous distillation

（4）Reflux ratio

2. Translate the following sentences into English or Chinese.

（1）Relative volatility can be described as the ratio of a particular compound to vaporize, relative to another compound.________________________________

____

（2）相对挥发度越大，分离就越容易。________________________________

____

小提示：

• the ratio of A to B A 与 B 的比值

• the greater…the easier 越大……越容易

Further Reading

Pressure Drop and Flooding

Plate columns (also commonly referred to as "tray columns") are essentially vertical cylinders in which the liquid and gas are contacted in stepwise fashion. The liquid enters at the top and flows downward *via* gravity. On the way, it flows across each plate and through a downspout to the plate below. The gas passes upward through openings of one sort or another in the plate, then bubbles through the liquid to form a froth, disengages from the froth, and passes on to the next plate above. The overall effect is a multiple, countercurrent contact of gas and liquid. Each plate of the column is a stage since the fluids on the plate are brought into intimate contact, interface diffusion occurs, and the fluids are separated. The number of theoretical plates (or stages) is dependent on the difficulty of the separation to be carried out and is determined solely from material balances, mass transfer resistances and equilibrium considerations. The diameter of the column, on the other hand, depends on the quantities of liquid and gas flowing through the column per unit time. The actual number of plates required for a given separation is greater than the theoretical number because of plate inefficiencies.

downspout *n.* 降液管

froth *n.* 泡沫

disengage *v.* 脱出

M12-19
单词读音及例句

semi-empirical *adj.* 半经验的

superficial *adj.* 表观的

conduit *n.* 导管

perforation *n.* 穿孔

weir *n.* 溢流堰

weeping *n.* 漏液（现象）

entrainment *n.* 雾沫夹带

likelihood *n.* 可能性

M12-20
单词读音及例句

There are numerous semi-empirical equations that are available for predicting the pressure drop across tray columns. As a preliminary estimate, one may assume the pressure drop is given by the height of liquid supported on the tray. Typically, this liquid height is in the 4 to 6 in range. Therefore, a reasonable approximation for pressure drop may be 4 to 6 in of H_2O per tray, which is approximately 0.1 to 0.2 psi per tray. The lower value applies to smaller diameter columns and the upper value applies to larger diameter units.

It is important that the vapor stream has the correct superficial velocity (linear average velocity, calculated as if the column conduit was empty) as it flows upwards in the column. Should the vapor flow too slowly, liquid may pass down through the tray perforations instead of over the weir, a condition known as weeping. However, if the vapor has a velocity that is too high, some liquid may be carried from the froth to the tray above by the rapidly flowing vapor. This condition is known as entrainment. Should the vapor velocity be increased further, entrainment may become excessive such that the liquid level in the downcomer will reach the plate above.

At this point, liquid flow from the tray(s) in question becomes hindered, and the column's entrainment flooding point has been reached. When calculating the allowable superficial velocity of a vapor, certain effects, such as the foaming tendency of a distillation mixture, are often taken into account, as foaming increases the likelihood of entrainment. Both excessive entrainment and weeping greatly influence tray efficiency and negatively impact overall column performance. (From Louis Theodore *etc.*, Unit Operations in Environmental Engineering, 2017, 658)

Reading comprehension

1. How do the gas and liquid contact in the plate column?
2. What are the effects of entrainment and weeping on overall column performance?

Lesson three: Extraction

In this lesson you will learn:

- Liquid-liquid extraction
- Extraction apparatus
- Solid-liquid extraction

1. Liquid-liquid extraction

Liquid-liquid extraction (or liquid extraction) is a process for separating a solute from a solution employing the concentration driving force between two immiscible (non-dissolving) liquid phases. Thus, liquid extraction involves the transfer of solute from one liquid phase into a second immiscible liquid phase.

A simple example is a familiar laboratory procedure: if an acetone-water solution is shaken in a separatory funnel with carbon tetrachloride and the liquids are allowed to settle, a large portion of the acetone will be found in the carbon-tetrachloride-rich phase and will thus have been separated from the water. A small amount of the water will also have been dissolved by the carbon tetrachloride, and a small amount of the latter will have entered the water layer, but these effects are relatively minor and can usually be neglected.

extraction *n.* 萃取

immiscible *adj.* 互不相溶的

acetone *n.* 丙酮

separatory funnel 分液漏斗

carbon tetrachloride 四氯化碳溶液

M12-21
单词读音及例句

Notes:

If an acetone-water solution is shaken in a separatory funnel with carbon tetrachloride and the liquids are allowed to settle, a large portion of the acetone will be found in the carbon-tetrachloride-rich phase and will thus have been separated from the water.

译文：如果在带有四氯化碳的分液漏斗中摇动丙酮-水溶液并静置，则大部分丙酮将存在于富含四氯化碳的相中，从而与水分离。

语法：句首 if 为连词，引导条件状语从句，从句中的并列连词 and 连接并列句；主句中 will... have been separated... 用完成时表示在将来某一时间以前已经完成的动作。

Liquid extraction is usually selected when distillation or stripping is impractical or too costly; this situation occurs when the relative volatility for two components falls between 1.0 and 1.2. Most organic solutes may be removed by this process. Extraction has been specifically used in removal and recovery of phenols, oils, and acetic acid from aqueous streams, and in removing and recovering freons and chlorinated hydrocarbons from organic streams.

The solution whose components are to be separated is the feed to the process. The feed is composed of a diluent and solute. The liquid contacting the feed for purposes of extraction is referred to as the solvent. If the solvent consists primarily of one substance (aside from small amounts of residual feed material that may be present in a recycled or recovered solvent), it is called a single solvent. A solvent consisting of a solution of one or more substances chosen to provide special properties is a mixed solvent. The solvent-lean, residual feed solution, with one or more constituents removed by extraction, is referred to as the raffinate. The solvent-rich solution containing the extracted solute(s) is the extract.

phenol *n.* 酚

chlorinated hydrocarbon 氯化烃

raffinate *n.* 萃余液

extract *n.* 提取物

M12-22
单词读音及例句

Notes:

The solvent-lean, residual feed solution, with one or more constituents removed by extraction, is referred to as the raffinate.

译文：通过萃取除去一种或多种成分的残留进料溶液——贫溶剂称为萃余液。

语法：句中 residual feed solution 为插入语；with… removed… 也为插入语，构成 with+*n.*+ 过去分词的复合宾语结构。

The degree of separation that arises because of the aforementioned solubility difference of the solute in the two phases may be obtained by providing multiple stage countercurrent contacting and subsequent separation of the phases, similar to a distillation operation. Recovery of the solute and solvent from the product stream is often carried out by stripping or distillation. The recovered solute may be either treated, reused, resold, or disposed of. Capital investment in this type of process primarily depends on the particular feed stream to be processed.

synthesis *n.* 合成

investigate *v.* 研究

dispersion *n.* 散布，分布

settling velocity 沉降速度

2. Extraction apparatus

There are two major categories of equipment for liquid extraction.

(1) Single-stage extraction

The first is single-stage units, which provide one stage of contact in a single device or combination of devices. In such equipment, the liquids are mixed, extraction occurs, and the insoluble liquids are allowed to separate as a result of their density differences. Several separate stages may be used in an application. They can readily be investigated in test apparatus. In particular, the following are to be tested:

M12-23
单词读音及例句

- The dispersion performance in the extraction.
- The settling velocity.
- Possible formation of mud can be caused by small amounts of impurities, which sometimes only become apparent after closure of the recycle streams. If necessary, the mud layer must be removed separately.

impurity *n.* 杂质

multistage extraction 多级萃取

incorporate *v.* 合并

cross current extraction 错流萃取

(2) Multistage extraction

Second, there are multistage devices, where many stages may be incorporated into a single unit. This type is normally employed in practice.

There are also two categories of operation: batch or continuous. In addition to cocurrent flow (rarely employed) provided in Figure 12-8 for a three-stage system, cross current extraction is a series of stages in which the raffinate from one extraction stage is contacted with additional fresh solvent in a subsequent stage. Cross current extraction is usually not economically appealing for

M12-24
单词读音及例句

large commercial processes because of the high solvent usage and low solute concentration in the extract.

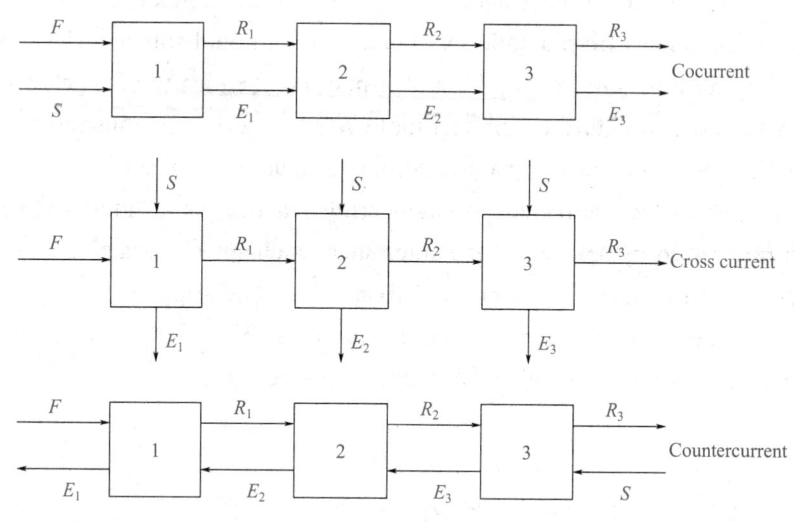

Figure 12-8　Multistage extractors

It can be shown that multistage cocurrent operation only increases the residence time and therefore will not increase the separation above that obtained in a single stage, provided equilibrium is established in a single stage. Crosscurrent contact, in which fresh solvent is added at each stage, will increase the separation beyond that obtainable in a single stage. However, it can also be shown that the degree of separation enhancement is not as great as can be obtained by countercurrent operation with a given amount of solvent.

residence time 停留时间

solvent　*n.* 溶剂

M12-25
单词读音及例句

Notes:

However, it can also be shown that the degree of separation enhancement is not as great as can be obtained by countercurrent operation with a given amount of solvent.

译文：然而，也可以看出，分离增强的程度并不如用一定量的溶剂逆流操作所能获得的那么大。

语法：句中 it 为先行主语，代表后移的由 that 引导的主语从句；not as… as 为用于比较的结构，意为“不像……那样……”。

The maximum separation that can be achieved between two solutes in a single equilibrium stage of the two phases is governed by equilibrium factors and the relative amounts of the two phases used, *i.e.*, the phase ratio. A combination of the overall and component mass balances with the equilibrium data allows the compositions of the phases at equilibrium to be computed. If the separation achieved is inadequate, it can be increased by either changing the phase ratio or by the addition of more contacting stages.

govern　*v.* 管理，支配

M12-26
单词读音及例句

leaching *n.* 浸出

particulate *adj.* 微粒的

cellular *adj.* 细胞的

permeable *adj.* 能透过的

lixiviation *n.* 浸滤

tannin *n.* 丹宁酸

3. Solid-liquid extraction

Solid-liquid extraction (leaching) involves the preferential removal of one or more components from a solid by contact with a liquid solvent. The soluble constituent may be solid or liquid, and it may be chemically or mechanically held in the pore structure of the insoluble solid material. The insoluble solid material is often particulate in nature, porous, cellular with selectively permeable cell walls, or surface-activated. In engineering practice, solid-liquid extraction is also referred to by several other names such as chemical extraction, washing extraction, diffusional extraction, and lixiviation. The simplest example of a leaching process is in the preparation of a cup of tea. Water is the solvent used to extract or leach tannins and other substances from the tea leaf.

M12-27
单词读音及例句

Reading comprehension

1. When to choose extraction instead of distillation to separate the two components?

2. How to improve the degree of separation by liquid extraction ?

3. Compare co-current, counter-current with cross-current extraction, which operation has the greatest separation enhancement with a given amount of solvent?

4. If the separation achieved is inadequate, what should we do to increase the separation?

Exercise

1. Noun explanation.

（1）Liquid-liquid extraction

（2）Raffinate

（3）Extract

2. Translate the following sentences into English or Chinese.

（1）A liquid-liquid extraction process may offer advantages in terms of higher selectivity or lower solvent usage and lower energy consumption, depending upon the application.____________________

____

（2）在为液体萃取过程选择溶剂时，有几个原则可以用作指导。

____

____

（3）液-液萃取也可以分批进行操作。____________________

____

小提示：
in terms of
在……方面
be carried out in…
进行，开展

Further Reading

Optimization of Solvent Consumption

Let us now consider the impact of the large number of purification steps on solvent consumption. Indeed, crystallization, liquid-liquid or liquid-solid extraction, preparative chromatography, and so on, are all associated with the use of solvents. A strategy on the choice of solvents and optimization of their use thus proves to be a key element. To address this challenge, several approaches are used. The first way is to set up a means for recycling the solvent, which is already quite a common practice in preparative chromatography. The quality control of the recycled solvent plays an essential role.

The problems associated with the consumption of solvent can also guide the selection process in advance. The resurgence of supercritical fluid chromatography (SFC), using a supercritical fluid as the mobile phase, is certainly related to its performance due to the low viscosity and high diffusivity of these media, but it is certainly also due to the ease of recycling this solvent. The development of crystallization processes, where supercritical CO_2 is used as an anti-solvent (SAS), helps, in addition to obtaining interesting properties of use, facilitates the removal of residual solvents. Similarly, extraction by supercritical CO_2 has the advantage of resulting in a spontaneous separation of substances extracted by simple depressurization.

Finally, a third approach aiming to reduce the environmental impact of solvents is the use of solvents from "green" chemistry and to minimize their consumption by increasing, for example, the extraction using microwave heating, ultrasound heating, and so on.

preparative *adj.* 预备的，准备的

chromatography *n.* 色谱分析法

resurgence *n.* 复活

supercritical fluid 超临界流体

diffusivity *n.* 扩散率

spontaneous *adj.* 自发的

depressurization *n.* 减压

M12-28
单词读音及例句

Reading comprehension

1. List three strategies that can reduce solvent consumption.
2. What are the advantages of extraction by supercritical CO_2?

Lesson four: Drying

In this lesson you will learn:

- Drying principles
- Factors affecting rate of drying
- Drying equipment

1. Drying principles

moisture *n.* 水分

simultaneous *adj.* 同时的

inert *adj.* 惰性的

vaporization *n.* 汽化

M12-29
单词读音及例句

The drying of solids to remove moisture involves the simultaneous processes of heat and mass transfer. Heat is transferred from the gas to the solid (and liquid) in order to evaporate the liquid contained in the solid. Mass is transferred as either a liquid or vapor within the solid and then as a vapor from the surface of the solid. The energy required to vaporize the liquid in a solid is almost always furnished by a hot, inert carrier gas that enters the drier. In some driers, the solid may be in contact with heated metal surfaces where the required heat of vaporization flows to the solid by conduction. In vacuum drying (where there is essentially no carrier gas), the heat of vaporization is furnished by conduction or radiation; here, the capacity of the drier is largely influenced by the heat-transfer surface available within the dryer.

Notes:

In some driers, the solid may be in contact with heated metal surfaces where the required heat of vaporization flows to the solid by conduction.

译文：在一些干燥器中，固体可能与加热的金属表面接触，在那里所需的汽化热通过热传导流向固体。

语法：句中关系副词 where 引导定语从句，修饰 heated metal surfaces。

sorption isotherm 吸附等温线

saturation *n.* 饱和度，饱和状态

capillary *n.* 毛细管

disposal *n.* 处理

sludge *n.* 污泥

M12-30
单词读音及例句

The sorption isotherm in a sorption balance allows an initial evaluation of the extent of the three drying steps:

- Surface liquid. Here the vapor pressure of the liquid corresponds to the saturation vapor pressure.
- Capillary-bound liquid. Here a lowering of the vapor pressure is expected for capillary diameters smaller than 0.1 mm.
- Dissolved or chemically bound liquid whose vapor pressure is established *via* the osmotic pressure or the chemical bonding.

Some drying operations involve only the exposure of solids to unsaturated air at normal temperature. Heat drying is expensive and is most often used where the dried material is of commercial value. In general, air drying is employed to facilitate the disposal of worthless sludge.

The curve provided in Figure 12-9 is obtained when a substance saturated with water is dried. During the drying process, a thin film of water exists on the surface of the material where water is supplied from the solid fast enough to keep the surface entirely wet. As this water is evaporated, water from the interior of the sample rises to the surface essentially by capillary action with the solid temperature approximately given by the wet-bulb temperature of the air. After drying has proceeded for some time, the surface film begins to disappear and the rate of drying decreases. This critical moisture content leads to dry patches on the surface, and the drying rate begins to fall. Ultimately, water ceases to evaporate and a final equilibrium moisture content (denoted by the dashed line in Figure 12-9) is achieved.

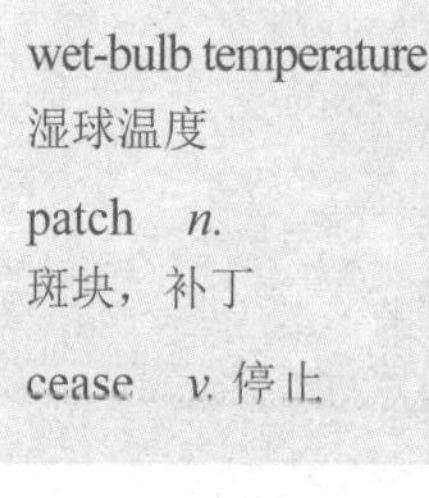

wet-bulb temperature 湿球温度

patch *n.* 斑块，补丁

cease *v.* 停止

M12-31
单词读音及例句

Figure 12-9 Typical drying process: moisture content *versus* time

Notes:

As this water is evaporated, water from the interior of the sample rises to the surface essentially by capillary action with the solid temperature approximately given by the wet-bulb temperature of the air.

译文：随着水的蒸发，样品内部的水主要通过毛细管作用上升到表面，直到固体温度大约等于空气的湿球温度。

语法：句首的 as 为连词，引导状语从句；approximately given by…the air 为过去分词短语作后置定语。

2. Factors affecting rate of drying

The air drying rate of materials depends on various process parameters, of which the most important are temperature, airflow and velocity of drying air, relative humidity, initial and final air and product temperature, and material size.

(1) Temperature

Temperature is the key element in a drying operation. The drying rate increases with temperature and the total drying time can be substantially reduced by using high temperature drying air, leading to enhanced capacity of the drying system. However, there seems to be a limit to which the drying air temperature can be increased, since excessively high temperatures may adversely affect quality with respect to discoloration, crack formation, and even burning of the dried product.

substantially *adv.* 实质上，大幅度

discoloration *n.* 污点，变色

M12-32
单词读音及例句

Notes:

However, there seems to be a limit to which the drying air temperature can be increased, since excessively high temperatures may adversely affect quality with respect to discoloration, crack formation, and even burning of the dried product.

译文：然而，干燥空气温度的升高似乎是有限度的，因为过高的温度可能会对干燥产品的质量产生不利影响，引起变色、形成裂纹甚至燃烧等。

语法：There seems to be... 句型，表示“似乎有”的意思；to which 引导定语从句，修饰 a limit；连词 since 引导状语从句，意为“因为”；with respect to 意为“关于，至于，就……而言”。

relative humidity 相对湿度

case hardening 表面硬化

sensible heat 显热

ambient *adj.* 环境的，外围的

saturate *v.* 浸透，使饱和

logarithmic *adj.* 对数的

M12-33 单词读音及例句

(2) Relative humidity

The drying rate increases, and drying time decreases, with reduction in relative humidity of the drying air, so wet air extends the drying time. However, a comparatively higher relative humidity of drying air is recommended in combination with high temperature because a high relative humidity assists in maintaining the capillary pores and thus minimizes the chances of case hardening.

(3) Air velocity

Air is usually used as a medium for heat and mass transfer in drying operations. Ideally, the air velocity (or flow rate) should be such that when the air exits the product bed, it should give up all its sensible heat and become saturated with moisture. If the air velocity is too low, the air temperature may drop rapidly to the ambient level and the air may become saturated with moisture before it reaches the top of the bed. If the hot air temperature falls below its dew point before exiting the bed, the moisture will condense in the upper layers of the bed. On the other hand, if a very high velocity of air is used for drying, its heat and mass transfer potential would be underutilized. At higher airflow rates, there is a logarithmic decrease in drying time.

(4) Size and shape

Reduction in size (length and thickness) increases the total surface area available for heat and mass transfer. The drying rate therefore increases with reduced size of material with consequent shorter drying time. Small-size materials may therefore be dried whole, whereas large ones should be cut into smaller pieces to achieve faster rates of drying, reduction in total drying time, and maintenance of quality.

Notes:

The drying rate therefore increases with reduced size of material with consequent shorter drying time.

译文：干燥速率随着材料尺寸的减小而增加，从而缩短干燥时间。

语法：句中包含两个介词 with 加名词短语作状语，表示“随着”。

3. Drying equipment

The drying process is carried out in one of the three basic dryer types.

The first is a continuous tunnel dryer (tray dryer) (Figure 12-10). In a continuous dryer, supporting trays with wet solids are move through an enclosed system while warm air blows over the trays. Trays of wet material loaded on trucks may be moved slowly through a drying tunnel. When a truck is dry, it is removed at one end of the tunnel, and a fresh one is introduced at the other end. Fresh air inlets and humid air outlets are spaced along the length of the tunnel to suit the rate of evaporation over the drying curve. This mode of operation is suited particularly to long drying times, from 20 to 96 h for the materials.

tunnel dryer 隧道式干燥器

enclosed system 封闭系统

M12-34
单词读音及例句

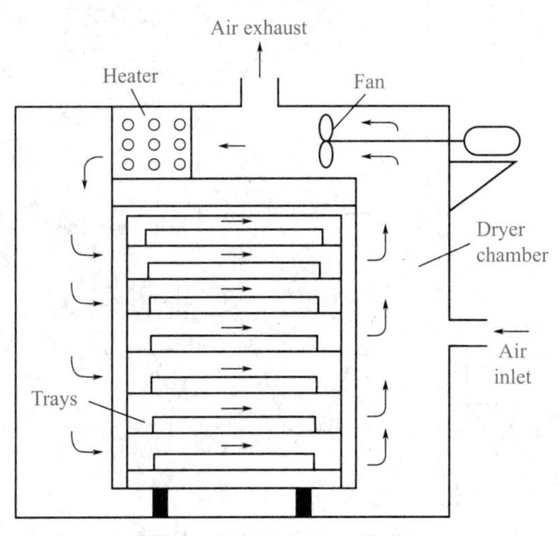

Figure 12-10　Tray dryer

Similar in concept to the continuous tunnel dryer, rotary dryers consist of an inclined rotating hollow cylinder (Figure 12-11). The rotary dryer is a popular device suitable for the drying of free-flowing materials that can be tumbled without concern for breaking. Moist solid is continuously fed into one end of a rotating cylinder with the simultaneous introduction of heated air. The cylinder is installed at a slight angle and with internal lifting flights so that the solid is showered through the hot air as it traverses the dryer. A rotary dryer is almost always operated in the cocurrent mode. The hot air is cooled as it is humidified; at the same time, the solid is heated and dried by contact with the hot air.

hollow *adj.* 中空的

rotary dryer 旋转干燥机

tumble *v.* 滚动，打滚

traverse *v.* 横穿

M12-35
单词读音及例句

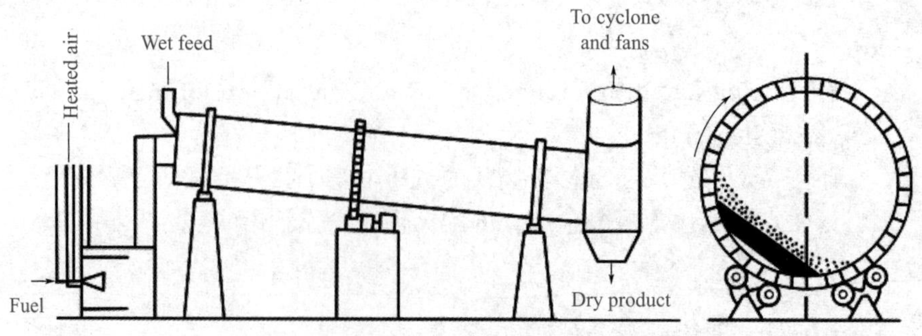

Figure 12-11　Rotary dryer

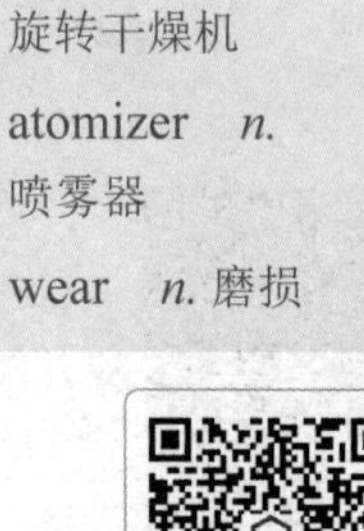
spray dryer
喷雾干燥器
rotary dryer
旋转干燥机
atomizer　*n.*
喷雾器
wear　*n.* 磨损

The final type of dryer is a spray dryer (Figure 12-12). In spray dryers, a liquid or slurry is sprayed through a nozzle, and fine droplets are dried by a hot gas, passed either concurrently or countercurrently, past the falling droplets. Spray dryers may be operated cocurrently or countercurrently. The action of the rotary atomizer results in smaller droplet size and size distribution, and is less subject to plugging and wear than the spray nozzle; however, it is higher in cost. This unit has found wide application in air pollution control.

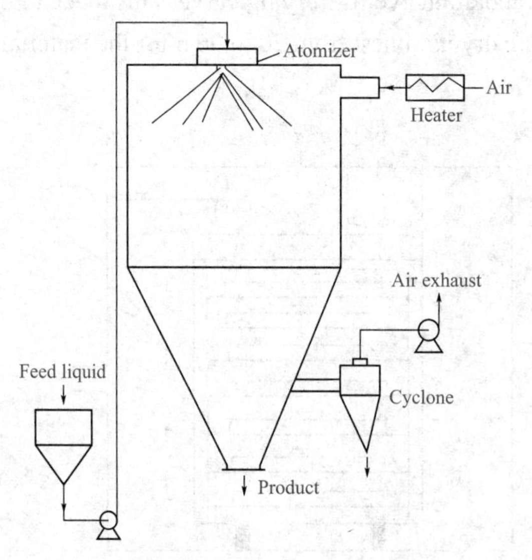

Figure 12-12　Spray dryer

Reading comprehension

1. What is drying?
2. Please describe the extent of the three drying steps.
3. Is the drying rate constant during the drying process?
4. What factors can affect the rate of drying?
5. List three basic dryer types.

Exercise

1. True or False.

(1) The drying rate increases with temperature, so the highest drying temperature should bc selected.

(2) A comparatively lower relative humidity of drying air is recommended in combination with high temperature.

(3) The rotary dryer is a popular device suitable for the drying of free-flowing materials that can be tumbled without concern for breaking.

2. Translate the following sentences into English or Chinese.

(1) Cocurrent drying leads to high drying rates and limits the maximum product temperatures, while countercurrent drying allows lower final moisture contents to be achieved.________________________________

____

(2) 应该注意的是，干燥是一种将液体与固体分离的方法。__________

____

小提示：

lead to… 导致

should be noted that…

应该注意……

Further Reading

Choice of Dryer

In the choice of a suitable dryer type, a series of aspects must be considered:

- Type of sorption isotherm, course of drying, final moisture content.
- Nature of the moisture (water, flammable solvent).
- Consistency of damp material (solvent, paste, crystal slurry, filter cake).
- Demands on the dry product (grain size and grain size distribution, freedom from dust, pourability, tendency to cake, rate of dissolution).
- Safety requirements (dust explosions).
- Permissible product temperature (formation of side products, melting).
- Tendency to form encrustations or highly viscous phases in the course of drying.

On the basis of the results of the sorption investigation and with consideration of the demands on the properties of the end product, a preliminary selection of dryer types can be made, the suitability of which is then checked experimentally. If the product is already in a given grain

flammable *adj.* 易燃的

consistency *n.* 稠度

grain *n.* 晶粒

pourability *n.* 流动性

permissible *adj.* 可允许的

encrustation *n.* 硬壳

M12-37
单词读音及例句

form, for example, resulting from a crystallization or precipitation, then various dryers of the contact or convective types can be chosen. For large capacities moisture the pneumatic dryer is particularly favorable with regard to investment cost. Fluidized-bed and flowing-bed dryers also allow longer drying times to be achieved. Blade dryers are suitable for a wide range of product properties.

The choice of dryer is more difficult when the drying step is also used to shape the product. Typical designs here are the spray dryer and the sprayed fluidized bed. If necessary, drying can be followed by a classification step in which undesired particles, for example, fines, are recycled to the feed stream for the dryer. When the desired product properties, for example, freedom from dust, pourability, rate of dissolution, and bulk density, are not attainable in the drying step, additional steps such as compaction and granulation must be used.

bulk density 容积密度
compaction *n.* 压紧，压实
granulation *n.* 造粒

M12-38
单词读音及例句

Convective drying is carried out where possible with air. The choice of co- or countercurrent drying is determined by the product properties. Cocurrent drying leads to high drying rates and limits the maximum product temperatures, while countercurrent drying allows lower final moisture contents to be achieved.

Reading comprehension

1. What additional steps should be used when the desired product properties, for example, freedom from dust, pourability, rate of dissolution, and bulk density, are not attainable in the drying step?

2. Whether co- or countercurrent drying should be chosen in convective drying?

项目五 绿色化工与安全生产

Part V: Green Chemistry and Safety Production

第十三单元　绿色化工与安全生产

Unit 13　Green Chemistry and Safety Production

Lesson one: Basis of Green Chemistry

In this lesson you will learn:

- Definition of green chemistry
- Green chemistry's 12 principles

1. Definition of green chemistry

Green chemistry is the design of chemical products and processes that reduce or eliminate the use or generation of hazardous substances. Green chemistry applies across the life cycle of a chemical product, including its design, manufacture, use, and ultimate disposal. Green chemistry is also known as sustainable chemistry.

The green chemistry approach seeks to redesign the materials that make up the basis of our society and our economy—including the materials that generate, store, and transport our energy—in ways that are benign for humans and the environment and possess intrinsic sustainability.

The concepts and practice of Green Chemistry have developed over nearly 20 years into a globe-spanning endeavor aimed at meeting the "triple bottom line"—sustainability in economic, social, and environmental performance.

The aphorism "an ounce of prevention is worth a pound of cure" is at the heart of Principle 1 of the Twelve Principles of Green Chemistry, a comprehensive set of design guidelines that have guided Green Chemistry development for many years.

eliminate *v.* 清除，消除

apply *v.* 使用，应用，实施

sustainable *adj.* 可持续的

benign *adj.* 和善的，良性的

M13-1 单词读音及例句

aphorism *n.* 格言，警句

stifle *v.* 压制，扼杀

M13-2 单词读音及例句

dispersal *n.* 分散，疏散，散布

grim *adj.* 令人不快的，令人沮丧

as a result of 为……的结果

innocuous *adj.* 无害的，无危险的

spontaneously *adv.* 自发地，自然地

tremendously *adv.* 特别，巨大，非常地

devise *v.* 发明，设计，想出

M13-3
单词读音及例句

The cost of handling, treating, and disposing hazardous chemicals is so high that it necessarily stifles innovation: funds must be diverted from research and development (scientific solutions) to hazard management (regulatory and political solutions, often).

Reviews of chemical accidents show that while the chemical industry is safer than other manufacturing jobs, exposure controls can fail. The consequence is injury and death to workers, which could have been avoided by working with less hazardous chemistry. Impacts on human health and the environment from the dispersal of hazardous waste are similarly grim, and monumental cleanup problems are faced as a result of the "treatment" rather than "prevention" approach.

In Green Chemistry, prevention is the approach to risk reduction: by minimizing the hazard portion of the equation, using innocuous chemicals and processes, risk cannot increase spontaneously through circumstantial means—accidents, spills, or disposal.

Green Chemistry has been tremendously successful in devising ways to reduce pollution through synthetic efficiency, catalysis, and improvements in solvent technology. Alternative synthetic methods have been applied to reduce energy consumption in the chemical industry, and bio-based feedstock is decreasing our reliance on depleted fossil resources.

Notes:

Green Chemistry has been tremendously successful in devising ways to reduce pollution through synthetic efficiency, catalysis, and improvements in solvent technology.

译文：绿色化学通过方法创新降低污染，在合成效率、催化作用和改进溶剂技术方面已经取得了巨大成功。

语法：句中 devising 为动名词，作介词宾语；to reduce 为不定式，作状语。

2. Green chemistry's 12 principles

breadth *n.* 宽度，广泛

depletable *adj.* 可耗减的

M13-4
单词读音及例句

The objective of green chemistry is to prevent waste formation rather than to treat it after it is formed by using chemical principles and new chemical technologies. In this regard, the formation of 12 well-known principles of green chemistry have been proposed. These principles demonstrate the breadth of the concept of green chemistry.

(1) Prevent waste

Design chemical syntheses to prevent waste. Leave no waste to treat or clean up.

(2) Maximize atom economy

Design syntheses so that the final product contains the maximum proportion of the starting materials. Waste few or no atoms.

(3) Design less hazardous chemical syntheses

Design syntheses to use and generate substances with little or no toxicity to either humans or the environment.

(4) Design safer chemicals and products

Design chemical products that are fully effective yet have little or no toxicity.

(5) Use safer solvents and reaction conditions

Avoid using solvents, separation agents, or other auxiliary chemicals. If you must use these chemicals, use safer ones.

(6) Increase energy efficiency

Run chemical reactions at room temperature and pressure whenever possible.

(7) Use renewable feedstock

Use starting materials (also known as feedstock) that are renewable rather than depletable. The source of renewable feedstock is often agricultural products or the wastes of other processes; the source of depletable feedstock is often fossil fuels (petroleum, natural gas, or coal) or mining operations.

(8) Avoid chemical derivatives

Avoid using blocking or protecting groups or any temporary modifications if possible. Derivatives use additional reagents and generate waste.

(9) Use catalysts, not stoichiometric reagents

Minimize waste by using catalytic reactions. Catalysts are effective in small amounts and can carry out a single reaction many times. They are preferable to stoichiometric reagents, which are used in excess and carry out a reaction only once.

(10) Design chemicals and products to degrade after use

Design chemical products to break down to innocuous substances after use so that they do not accumulate in the environment.

(11) Analyze in real time to prevent pollution

Include in-process, real-time monitoring and control during syntheses to minimize or eliminate the formation of byproducts.

(12) Minimize the potential for accidents

Design chemicals and their physical forms (solid, liquid, or gas) to minimize the potential for chemical accidents including explosions, fires, and releases to the environment. These principles are shown in Figure 13-1.

Since green chemistry was established, there have been tremendous advances in the industry. Nevertheless, there remains considerable room for improvement. The chemical industry faces a number of significant challenges, from reducing its dependence on fossil fuels to playing its part in addressing climate change more generally.

Specific challenges include: capturing and fixing carbon dioxide and other greenhouse gases; developing a greater range of biodegradable plastics; reducing the high levels of waste in pharmaceutical drug manufacture; and improving the efficiency of water-splitting by employing visible light photocatalysts.

derivative *n.* 衍生物

catalyst *n.* 催化剂

stoichiometric reagent 化学计量试剂

degrade *v.* 降解，分解

real time 实时

M13-5 单词读音及例句

tremendous *adj.* 巨大的，极大的

climate change 气候变化

biodegradable *adj.* 可生物降解的

pharmaceutical *adj.* 制药的，配药的

human capital 人力资本

M13-6 单词读音及例句

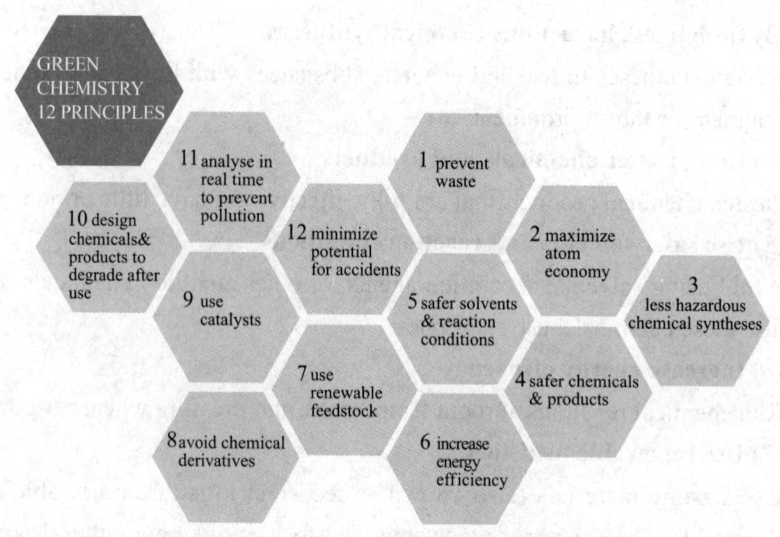

Figure 13-1　Green chemistry's 12 principles

History suggests that society can develop creative solutions to complex, intractable problems. However, success will most likely require a concerted approach across all areas of science, strong leadership, and a willingness to strategically invest in human capital and value fundamental research.

Notes:

However, success will most likely require a concerted approach across all areas of science, strong leadership, and a willingness to strategically invest in human capital and value fundamental research.

译文：然而，成功很可能需要在所有科学领域采取协调一致的方法，强有力的领导，并愿意战略性地投资于人力资本和重视基础研究。

语法：需分清句中两个并列连词 and 连接的成分：第一个 and 连接 a concerted approach…，strong leadership 和 a willingness to…；第二个 and 连接 to invest in… 和 value…。

Reading comprehension

1. What is the "triple bottom line" of Green Chemistry?

2. The 12 well-known principles involve the various aspects of the synthetic chemistry and the chemical processes, can you list some of them?

Exercise

Translate the following sentences into English or Chinese.

(1) Green chemistry is the design of chemical products and processes that

reduce or eliminate the use or generation of hazardous substances.____________

____

（2）一分的预防胜于十分的治疗。__________________________

____

Further Reading

A Secret to Cleaner Chemistry

Catalysts are substances that accelerate reactions, typically by enabling chemical bonds to be broken and/or formed without being consumed in the process. Not only do they speed up reactions, but they can also facilitate chemical transformations that might not otherwise occur.

In principle, only a very small quantity of a catalyst is needed to generate copious amounts of a product, with reduced levels of waste.

The development of new catalytic reactions is one particularly important area of green chemistry. As well as being more environmentally friendly, these processes are also typically more cost effective.

Catalysts take many forms, including biological enzymes, small organic molecules, metals, and particles that provide a better surface for reactions to take place. Roughly 90% of industrial chemical processes use catalysts and at least 15 Nobel Prizes have been awarded for catalysis research. This represents a tremendously important and active area of both fundamental and applied research.

catalyst
n. 催化剂，触媒剂

not only…but also
不但……而且

cost effective
节省成本有效果的

as well as
既……又……，此外

M13-7
单词读音及例句

Reading comprehension

1. What are catalysts?

2. Roughly 90% of industrial chemical processes use catalysts and at least 15 Nobel Prizes have been awarded for catalysis research. What does that mean?

Lesson two: New Conversion Methods

In this lesson you will learn:

- Plasma technology
- Microwave technology
- Ultrasound technology

Many nontraditional technologies, such as plasma, microwave, ultrasonic, and radiation technology, use nonthermal energy to intensify the process.

1. Plasma technology

conversion *n.* 转变，转换，转化

plasma *n.* 血浆，等离子体

M13-8
单词读音及例句

Plasma technology is based on a simple physical principle. Matter changes its state when energy is supplied to it as shown in Figure 13-2: solids become liquid, and liquids become gaseous. If even more energy is supplied to a gas, it is ionized and goes into the energy-rich plasma state, the fourth state of matter.

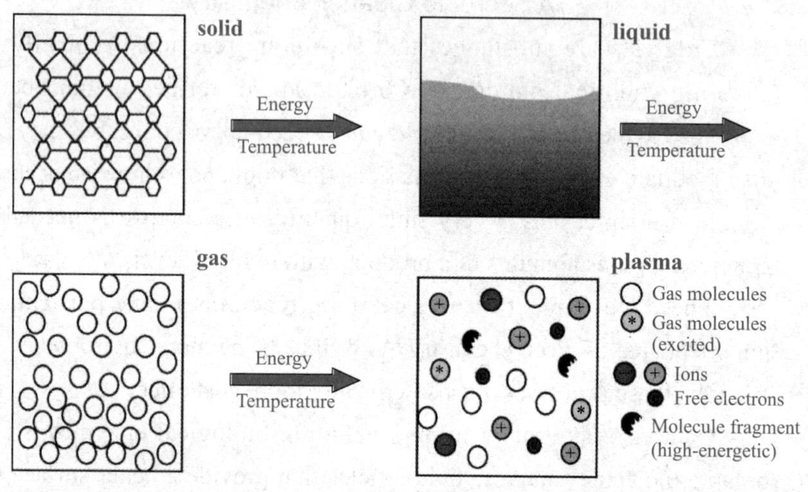

Figure 13-2 Matter changes its state when energy is supplied to it

Arctic *n.* 北极地区

Antarctic *n.* 南极地区

eclipse *n.* 日食，月食

regardless of *prep.* 不管，不顾

charged particle 带电粒子

neutral *adj.* 中立的，中性的

electron temperature 电子温度

M13-9
单词读音及例句

With increasing energy input, the state of matter changes from solid to liquid to gaseous. If additional energy is then fed into a gas by means of electrical discharge, the gas will turn into plasma.

Plasma was first discovered by Irving Langmuir in 1928. It is not rare; actually, quite the opposite is true. More than 99% of the visible matter in the universe is in the plasma state. It can be seen in its natural form on earth as lightning or as polar light in the Arctic and Antarctic, for example. During a solar eclipse, plasma can be observed as a bright circle of light (corona) around the sun.

Regardless of partially ionized or fully ionized gas, the total number of positive charge and negative charge on the numerical is always equal; in the macro the charge is neutral, called plasma. The plasma can be divided into two categories based on the relative height of the charged particle energy in the plasma (usually expressed as electron temperature). One is the high-temperature plasma, that is, the electron temperature in the tens of electron volts (1eV amounts to 11, 600 K) above the plasma. The other is the low-temperature plasma, that is, the electron temperature in the plasma of tens of electron volts. Low-temperature plasma has been widely used in many fields, such as materials, information, energy, chemical, metallurgy, machinery, military and aerospace.

Notes:

It is not rare; actually, quite the opposite is true. More than 99% of the visible matter in the universe is in the plasma state.

译文：这并不罕见；事实上，恰恰相反。宇宙中 99% 以上的可见物质处于等离子体状态。

语法：英语中分号可用来连接两个或多个独立的句子，第一句话中两个独立的句子的语义关系紧密，用分号连接可避免结构松散。百分比 percent（或 per cent, %）作主语时的主谓一致，百分比加不可数名词或单数名词时，谓语动词一般用单数形式。

2. Microwave technology

Microwaves are a type of electromagnetic radiation, as are radio waves, ultraviolet radiation, X-rays and gamma-rays. Microwaves have a range of applications, including communications, radar and, perhaps best known by most people, cooking.

Electromagnetic radiation is transmitted in waves or particles at different wavelengths and frequencies. This broad range of wavelengths is known as the electromagnetic spectrum (EM spectrum). The spectrum is generally divided into seven regions in order of decreasing wavelength and increasing energy and frequency. The common designations are radio waves, microwaves, infrared (IR), visible light, ultraviolet (UV), X-rays and gamma-rays. Microwaves fall in the range of the EM spectrum between radio and infrared light.

Figure 13-3 illustrates that microwaves have frequencies ranging from about 1 billion cycles per second, or 1 gigahertz (GHz), up to about 300 gigahertz and wavelengths of about 30 centimeters (12 inches) to 1 millimeter (0.04 inches). This region is further divided into a number of bands, with designations such as L, S, C, X and K.

a range of
一系列，一排

in order of
按……顺序

designation *n.*
任命，名称，称号

M13-10
单词读音及例句

ELECTROMAGNETIC SPECTRUM

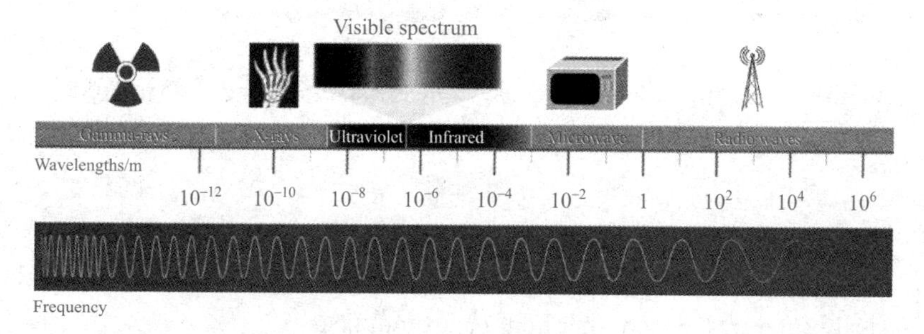

Figure 13-3 Frequencies and wavelengths ranging of microwaves

accelerate *v.* 加速，加快

conventional *adj.* 依照惯例的，传统的

incineration *n.* 焚烧，焚化

dioxin *n.* 二噁英，二氧（杂）芑

toxin *n.* 毒素

M13-11
单词读音及例句

Microwave technology can accelerate the rate of chemical reaction, change the course of the chemical reaction, obtain a new reaction product and realize the reaction which cannot be carried out by some conventional methods. At present, the microwave-assisted synthesis has been successfully applied in many responses. Study of microwave technology in the synthesis of inorganic materials, such as hard alloys, high-temperature materials, ceramic materials, nanomaterials, metal compounds, synthetic diamond and so on, is very extensive, and has made good progress in many ways. A large number of medical wastes are produced every year in the world, which caused serious environmental pollution. If the microwave technology is used, more than 60% of the medical waste can be used as landfill. Compared with the traditional incineration method, it will not produce dioxin with strong toxin and secondary pollutants, and has the characteristics of fast processing speed, good effect, low energy consumption and so on.

3. Ultrasound technology

Ultrasound is sound waves with frequencies higher than the upper audible limit of human hearing. Ultrasound is not different from "normal" (audible) sound in its physical properties, except that humans cannot hear it. This limit varies from person to person and is approximately 20 kilohertz (20,000 hertz) in healthy young adults. Ultrasound devices operate with frequencies from 20 kHz up to several gigahertz.

ultrasound *n.* 超声，超声波

audible *adj.* 听得见的

infrasound *n.* 次声

acoustic *adj.* 声音的，听觉的

Ultrasound is used in many different fields. Ultrasonic devices are used to detect objects and measure distances. Ultrasound imaging or sonography is often used in medicine. In the nondestructive testing of products and structures, ultrasound is used to detect invisible flaws. Industrially, ultrasound is used for cleaning, mixing, and accelerating chemical processes. Animals such as bats and porpoises use ultrasound for locating prey and obstacles. Figure 13-4 illustrates the approximate frequency ranges corresponding to ultrasound, with rough guide of some applications.

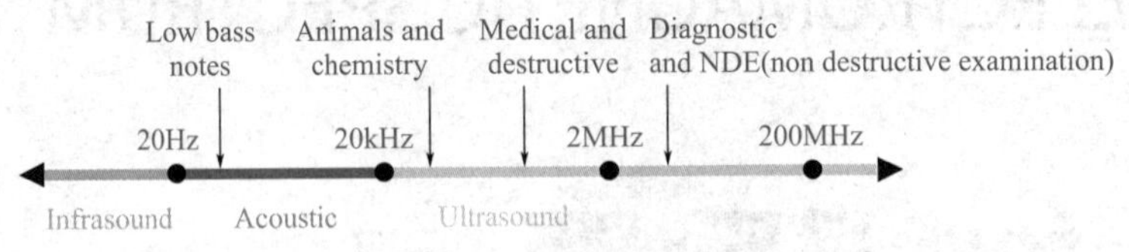

Figure 13-4 Approximate frequency ranges corresponding to ultrasound, with rough guide of some applications

M13-12
单词读音及例句

Notes:

Ultrasound is not different from "normal" (audible) sound in its physical properties, except that humans cannot hear it.

译文：超声波在物理性质上与“正常”（可听）声音没有区别，只是人类听不到。

语法：句中 except that 引导状语从句，意为“除了，只是”。

Reading comprehension

1. Why in the macro the plasma is neutral?
2. How do we describe the energy of plasma particles?
3. How is electromagnetic radiation transmitted?

Exercise

1. True or false.

(1) Plasma is the fourth state of matter.

(2) Compared with the traditional incineration method, microwave technology will not produce dioxin with strong toxin and secondary pollutants.

(3) We can hear ultrasound.

2. Translate the following sentences into English or Chinese.

(1) Compared with the traditional incineration method, it will not produce dioxin with strong toxin and secondary pollutants, and has the characteristics of fast processing speed, good effect, low energy consumption and so on.

____

____

(2) 光谱一般按波长递减、能量和频率递增的顺序分为七个区域。

____

____

Further Reading

Ultrasonic Range Finding

A common use of ultrasound is in underwater range finding; this use is also called sonar. Figure 13-5 shows the principle of an active sonar. An ultrasonic pulse is generated in a particular direction. If there is an object in the path of this pulse, part or all of the pulse will be reflected back to the transmitter as an echo and can be detected through the receiver path. By measuring the difference in time between the pulse being transmitted and the echo being received, it is possible to determine the distance.

underwater range finding　水下测距

sonar　*n.* 声呐

difference in time　时差

M13-13
单词读音及例句

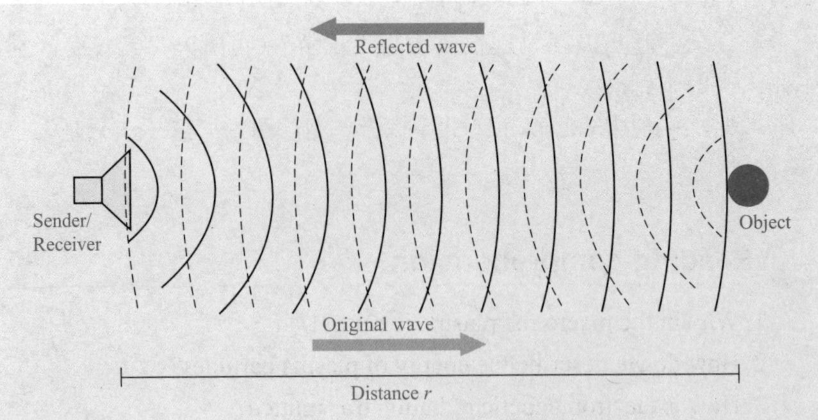

Figure 13-5　Principle of an active sonar

The measured travel time of sonar pulses in water is strongly dependent on the temperature and the salinity of the water. Ultrasonic ranging is also applied for measurement in air and for short distances. For example, hand-held ultrasonic measuring tools can rapidly measure the layout of rooms.

Although range finding underwater is performed at both sub-audible and audible frequencies for great distances (1 to several kilometers), ultrasonic range finding is used when distances are shorter and the accuracy of the distance measurement is desired to be finer. Ultrasonic measurements may be limited through barrier layers with large salinity, temperature or vortex differentials. Ranging in water varies from about hundreds to thousands of meters, but can be performed with centimeters to meters accuracy.

Reading comprehension

1. What is the working principle of sonar?
2. What is the advantage of ultrasonic sonar in range finding underwater?

Lesson three: Green Catalysts and Green Solvents

In this lesson you will learn:

- Green catalysts
- Green solvents

1. Green catalysts

(1) Introduction

In chemistry, catalysts refer to changing the speed of a reaction using a substance that's not consumed by the reaction.

Catalysts are used for many reactions that would otherwise be very slow, or would not happen at all. In the presence of suitable catalysts, these processes unfold very quickly.

In 2005, the Nobel Prize in chemistry was awarded for the discovery of a catalytic chemical process called metathesis. It has broad applicability in the chemical industry yet uses significantly less energy and could reduce greenhouse gas emissions for many key processes. The process is stable at normal temperatures and pressures, which can be used in combination with greener solvents, and it is likely to produce less hazardous waste.

Without catalysts, humans wouldn't have got far. They trigger many of the processes on which we rely, from age-old mechanisms deep down in our cells to the production of contemporary consumer goods. They could also smooth our path towards a sustainable future.

(2) Focus principles

Reduce derivatives—a selective catalytic or enzymatic pathway to a chemical can use fewer reagents and energy while providing a more direct synthetic route (compared to stoichiometric syntheses).

Catalysis—selective catalytic reagents are superior to stoichiometric ones; they can minimize or eliminate waste, depending on the reaction, which is necessary to achieve greener chemistry.

Maximize resource efficiency—catalysis can offer more direct synthetic routes, optimizing the time, energy and materials required to make a desired product.

in the presence of 在……面前

Nobel Prize 诺贝尔奖

unfold *v.* 展开，打开

metathesis *n.* 复分解

M13-14 单词读音及例句

greenhouse gas 温室气体

trigger *v.* 触发

age-old *adj.* 古老的

contemporary *adj.* 当代的

M13-15 单词读音及例句

Notes:

Reduce derivatives—a selective catalytic or enzymatic pathway to a chemical can use fewer reagents and energy while providing a more direct synthetic route (compared to stoichiometric syntheses).

译文：还原衍生物——一种选择性催化或酶促合成化学品的途径，可以使用更少的试剂和能量，同时提供更直接的合成路线（与化学计量合成相比）。

语法：while 后面接状语从句，意为“当……的时候”，括号中的部分为过去分词短语作状语。

affordable *adj.* 买得起的

eagerly awaited 翘首以待

M13-16
单词读音及例句

(3) Example of solid catalysts

And indeed, different processes require different catalysts, many of which are based on expensive and rare metals such as ruthenium, palladium or platinum. Finding more common and affordable substances to set these processes in motion would help to make them less costly and run them sustainably on a much larger scale.

This possibility could contribute to eagerly awaited technological breakthroughs—such as the development of an inexpensive, environmentally friendly process for the production of hydrogen. As a clean and potentially abundant energy carrier, this substance could help to reduce our reliance on fossil fuels.

One research team was determined to make its contribution to the advent of the hydrogen economy. Instead of the platinum catalysts that predominate in hydrogen production, it uses molybdenum, combined with sulphur. Molybdenum is about a thousand times less expensive than platinum, and a thousand times more abundant.

extract *vt.* 提取

exploitable *adj.* 可利用的

M13-17
单词读音及例句

This catalyst is intended for a solar-powered process that extracts hydrogen from water as shown in Figure 13-6. While many other technologies have to come together to transform this approach into an exploitable technology, it has already achieved very promising results at the lab scale. It is thus paving the way towards affordable and sustainable production, conversion, storage and transport of solar energy.

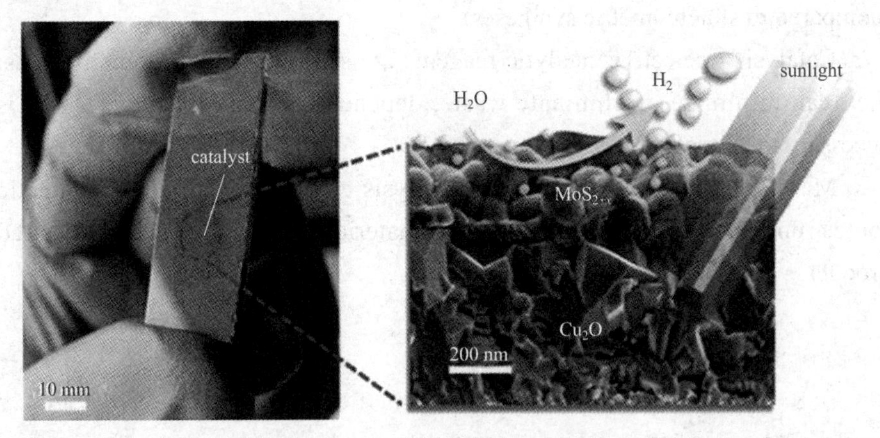

Figure 13-6 A layered device for sunlight-driven production of hydrogen from water

Notes:

Molybdenum is about a thousand times less expensive than platinum, and a thousand times more abundant.

译文：钼的价格大约比铂便宜一千倍，含量也比铂丰富一千倍。

语法：英文倍数比较的写法为“基数词 +times（三倍和三倍以上）+ more/less than…”。

2. Green solvents

(1) Introduction

The major application of solvents in human activities is in paints and coatings (46% of usage). Smaller volume applications include cleaning, de-greasing, adhesives, and in chemical synthesis. Traditional solvents are often toxic or are chlorinated. Green solvents, on the other hand, are generally less harmful to health and the environment and preferably more sustainable.

innocuous *adj.* 无害的

enclosed *adj.* 封闭的

feasible *adj.* 可行的

M13-18
单词读音及例句

Ideally, solvents would be derived from renewable resources and biodegrade to innocuous, often a naturally occurring product. However, the manufacture of solvents from biomass can be more harmful to the environment than making the same solvents from fossil fuels. Thus the environmental impact of solvent manufacture must be considered when a solvent is being selected for a product or process. Another factor to consider is the fate of the solvent after use. If the solvent is being used in an enclosed situation where solvent collection and recycling are feasible, then the energy cost and environmental harm associated with recycling should be considered; in such a situation water, which is energy-intensive to purify, may not be the greenest choice. On the other hand, a solvent contained in a consumer product is likely to be released into the environment upon use, and therefore the environmental impact of the solvent itself is more important than the energy cost and impact of solvent recycling; in such a case water is very likely to be a green choice.

In short, the impact of the entire lifetime of the solvent, from cradle to grave (or cradle to cradle if recycled) must be considered. Thus the most comprehensive definition of a green solvent is the following: "A green solvent is the solvent that makes a product or process has the least environmental impact over its entire life cycle."

Owing to their special properties, green solvents improve chemical processes, lower the use of solvents and decrease the processing steps. Water, supercritical CO_2, ionic liquids, non-toxic liquid polymers and their diverse combinations are part of the class of green solvents as shown in Figure 13-7. They are characterized by low toxicity, convenient accessibility and the possibility of reuse as well as great efficiency. An ideal green solvent would also mediate reactions, separations or catalyst recycling. An idea of green chemistry is aiming for replacement of commonly used solvents with 'green' ones, resulting in a reduced environmental impact.

supercritical *adj.* 超临界的

mediate *v.* 调停，调解

M13-19
单词读音及例句

(2) Example of green solvents

Supercritical carbon dioxide (sCO_2) is a fluid state of carbon dioxide where it is held at or above its critical temperature and critical pressure.

Carbon dioxide usually behaves as a gas in air at standard temperature and pressure (STP), or as a solid called dry ice when cooled and/or pressurised

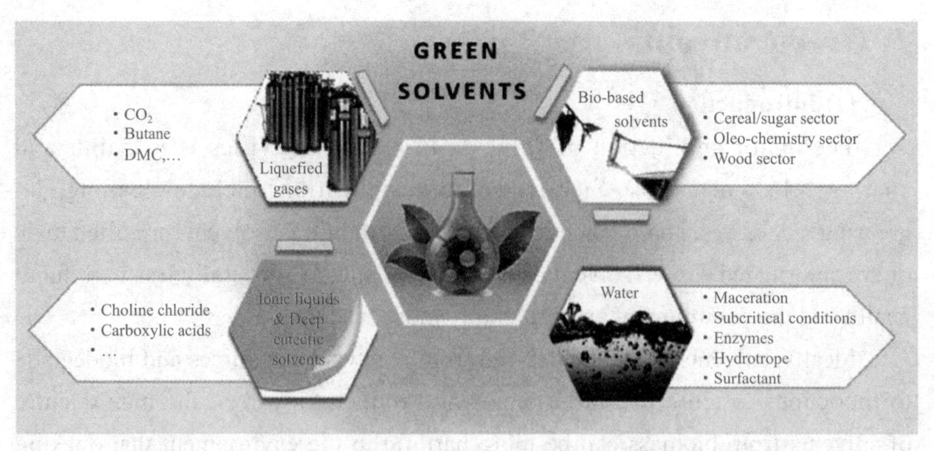

Figure 13-7 Current opinion in green and sustainable chemistry

sufficiently. If the temperature and pressure are both increased from STP to be at or above the critical point for carbon dioxide as shown in Figure 13-8, it can adopt properties midway between a gas and a liquid. More specifically, it behaves as a supercritical fluid above its critical temperature (304.13 K, 31.0 ℃, 87.8 ℉) and critical pressure (7.3773 MPa, 72.8 atm, 1,070 psi, 73.8 bar), expanding to fill its container like a gas but with a density like that of a liquid.

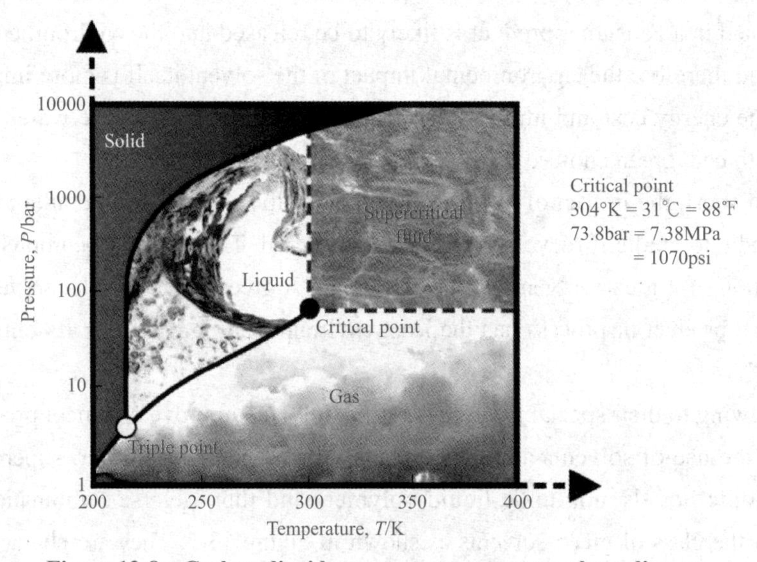

Figure 13-8 Carbon dioxide pressure-temperature phase diagram

Supercritical CO_2 is becoming an important commercial and industrial solvent due to its role in chemical extraction in addition to its low toxicity and environmental impact. The relatively low temperature of the process and the stability of CO_2 also allow most compounds to be extracted with little damage or denaturing. In addition, the solubility of many extracted compounds in CO_2 varies with pressure, permitting selective extractions.

Supercritical carbon dioxide is used as the extraction solvent for creation of essential oils and other herbal distillates. Its main advantages over solvents such

denaturing *n.* 变性

solubility *n.* 溶（解）度

essential oil 精油

herbal distillate 草药提取物

M13-20
单词读音及例句

as hexane and acetone in this process are that it is non-toxic and non-flammable. Furthermore, separation of the reaction components from the starting material is much simpler than with traditional organic solvents. The CO_2 can evaporate into the air or be recycled by condensation into a cold recovery vessel. Its advantage over steam distillation is that it operates at a lower temperature, which can separate the plant waxes from the oils.

Notes:

Its advantage over steam distillation is that it operates at a lower temperature, which can separate the plant waxes from the oils.

译文：与水蒸气蒸馏相比，它的优点是操作温度较低，可以将植物蜡从油中分离出来。

语法：句中 that 为连词，引导表语从句；which 为关系代词，引导非限制性定语从句。

Reading comprehension

1. What is supercritical carbon dioxide (sCO_2)?

2. What are the advantages of supercritical carbon dioxide as an extraction solvent?

Exercise

1. True or false.

（1）Catalysts can slow down chemical reactions.

（2）Catalysis can minimize or eliminate waste, depending on the reaction, which is necessary to achieve greener chemistry.

（3）Water is sometimes a green solvent and sometimes not, depending on the situation.

2. Translate the following sentences into Chinese.

（1）It has broad applicability in the chemical industry yet uses significantly less energy and could reduce greenhouse gas emissions for many key processes.

____

____

（2）In short, the impact of the entire lifetime of the solvent, from cradle to grave (or cradle to cradle if recycled) must be considered.

____

____

Further Reading

near-ambient 近环境

lattice *n.* 晶格

M13-21
单词读音及例句

Ionic liquid

An ionic liquid (IL) is a salt in the liquid state. In some contexts, the term has been restricted to salts whose melting point is below some arbitrary temperature, such as 100 ℃ (212 ℉). While ordinary liquids such as water and gasoline are predominantly made of electrically neutral molecules, ionic liquids are largely made of ions. These substances are variously called liquid electrolytes, ionic melts, ionic fluids, fused salts, liquid salts, or ionic glasses.

Ionic liquids have many potential applications. They are powerful solvents and can be used as electrolytes. Salts that are liquid at near-ambient temperature are important for electric battery applications, and have been considered as sealants due to their very low vapor pressure.

Any salt that melts without decomposing or vaporizing usually yields an ionic liquid. Sodium chloride (NaCl), for example, melts at 801 ℃ (1,474 ℉) into a liquid that consists largely of sodium cations (Na^+) and chloride anions (Cl^-). Conversely, when an ionic liquid is cooled, it often forms an ionic solid—which may be either crystalline or glassy.

The ionic bond is usually stronger than the van der Waals forces between the molecules of ordinary liquids. Because of these strong interactions, salts tend to have high lattice energies, manifested in high melting points. Some salts, especially those with organic cations, have low lattice energies and thus are liquid at or below room temperature.

Reading comprehension

1. What is the difference between ionic liquids and ordinary liquids?
2. Please list two potential applications of ionic liquids.

Lesson four: Chemical Safety Technology and Management

In this lesson you will learn:

- Chemical safety technology
- Safety management

1. Chemical safety technology

Chemicals as elements, compounds, mixtures, solutions and emulsions are very widely used and transported in the modern industrial society. Of necessity, they are also used in schools, universities and other training facilities to educate pupils in their safe use and handling and also are commonly used in domestic situations for cleaning, garden maintenance and DIY.

emulsion *n.* 乳浊液
pupil *n.* 学生
ecosystem *n.* 生态系统

Chemical safety includes all those policies, procedures and practices designed to minimize the risk of exposure to potentially hazardous chemicals. This includes the risks of exposure to persons handling the chemicals, to the surrounding environment, and to the communities and ecosystems within that environment.

M13-22
单词读音及例句

The hazardous nature of many chemicals may be increased when mixed with other chemicals, heated or handled inappropriately. In a chemically safe environment, users are able to take appropriate actions in case of accidents although many incidents of exposure to chemical hazards occur outside of controlled environments such as manufacturing plants or laboratories.

2. Chemical safety management

The management and control of chemical safety is widely developed through primary legislation and orders derived from such legislation. A person conducting a business or undertaking must manage risks associated with using, handling, generating or storing of hazardous chemicals at a workplace. In order to manage risk, a duty holder must undertake the following tasks:

(1) Identifying hazards

The first step in managing risks involves identifying all the chemicals that are used, handled, stored or generated at your workplace in consultation with workers. The identity of chemicals in the workplace can usually be determined by looking at the label and the safety data sheets (SDS) as shown in Figure 13-9, and reading what ingredients are in each chemical or product. In some cases, a chemical may not have a label or an SDS, for example where fumes are generated in the workplace from an activity such as welding.

in consultation with 与……磋商
fume *n.* 烟，气
welding *n.* 焊接

M13-23
单词读音及例句

A manufacturer or importer must determine the hazards of a chemical against specified criteria. This process is known as classification, and it is the hazard classification of a chemical that determines what information must be included on labels and SDS, including the type of label elements, hazard statements and pictograms. Manufacturers and importers are required to provide labels and SDS, and must review the information on them at least once every five years or whenever necessary to ensure the information contained in the SDS is correct, for example new information on a chemical may lead to a change in its hazard classification.

SAFETY DATA SHEET

Hydrogen Sulfide

Section 1. Identification

GHS product identifier	: Hydrogen Sulfide
Chemical name	: hydrogen sulphide
Other means of identification	: Hydrogen sulfide; Hydrogen sulfide (H_2S); Sulfuretted hydrogen; Sewer gas; Hydrosulfuric acid; dihydrogen sulfide
Product use	: Synthetic/Analytical chemistry.
Synonym	: Hydrogen sulfide; Hydrogen sulfide (H_2S); Sulfuretted hydrogen; Sewer gas; Hydrosulfuric acid; dihydrogen sulfide
SDS #	: 001029
Supplier's details	:
Emergency telephone number (with hours of operation)	:

Section 2. Hazards identification

OSHA/HCS status	: This material is considered hazardous by the OSHA Hazard Communication Standard (29 CFR 1910.1200).
Classification of the substance or mixture	: FLAMMABLE GASES - Category 1 GASES UNDER PRESSURE - Liquefied gas ACUTE TOXICITY (inhalation) - Category 2 SPECIFIC TARGET ORGAN TOXICITY (SINGLE EXPOSURE) (Respiratory tract irritation) - Category 3 AQUATIC HAZARD (ACUTE) - Category 1

GHS label elements

Hazard pictograms	:
Signal word	: Danger
Hazard statements	: Extremely flammable gas. May form explosive mixtures with air. Contains gas under pressure; may explode if heated. May cause frostbite. Fatal if inhaled. Extended exposure to gas reduces the ability to smell sulfides. May cause respiratory irritation. Very toxic to aquatic life.

Precautionary statements

Figure 13-9　An example of SDS

assess　*vt.* 评估，评定

carcinogen　*n.* 致癌物

mutagen　*n.* 诱变剂

M13-24 单词读音及例句

(2) Assessing risks

The type of risk assessment that should be conducted will depend on the nature of the work being performed.

① A basic assessment consists of: reviewing the label and the SDS of the hazardous chemicals and assessing the risks involved in their use.

Deciding whether the hazardous chemicals in the workplace are already controlled with existing control measures, as recommended in the SDS or other reliable sources, or whether further control measures are needed.

② In a generic assessment, an assessment is made of a particular workplace, area, job or task and the assessment is then applied to similar work activities that involve the use of the chemical being assessed.

③ A detailed assessment may be needed when there is a significant risk to health and for very high risk chemicals such as carcinogens, mutagens, reproductive toxicants or sensitization agents in the case of health hazards. Information on the label and SDS will allow you to determine whether the chemical has these hazards. A more detailed assessment may also be required when there is uncertainty as to the risk of exposure or health.

In order to complete a detailed assessment, further information may be sought and decisions should be taken to:

eliminate the uncertainty of any risks,

select appropriate control measures,

ensure that control measures are properly used and maintained, and determine if air monitoring or health monitoring are required.

It may be necessary to engage external professional assistance to undertake a more detailed assessment.

(3) Controlling risks

There are a number of ways to control the risks associated with hazardous chemicals. Some control measures are more effective than others. Control measures can be ranked from the highest level of protection and reliability to the lowest. This ranking is known as the hierarchy of control.

You must always aim to eliminate a hazard and associated risk first. If this is not reasonably practicable, the risk must be minimised by using one or more of the following approaches:

substitution

isolation

implementing engineering controls

If a risk then remains, it must be minimised by implementing administrative controls, so far as is reasonably practicable. Any remaining risk must be minimised with suitable personal protective equipment (PPE).

Administrative control measures and PPE rely on human behaviour and supervision and when used on their own, tend to be the least effective ways of minimising risks.

(4) Monitoring and review

Health monitoring of a person means monitoring the person to identify changes in the person's health status because of exposure to certain substances. It involves the collection of data in order to evaluate the effects of exposure and to confirm that the absorbed dose is within safe levels. This allows decisions to be made about implementing ways to eliminate or minimise the worker's risk of exposure.

Control measures must be reviewed (and revised if necessary). When reviewing the control measures, consultation must occur with workers and their health and safety representatives.

substitution *n.* 取代，替换

monitor *vt.& n.* 检测

revise *v.* 修订

consultation *n.* 咨询，商讨

M13-25
单词读音及例句

(5) Emergency preparedness

Regardless of controls put in place to prevent incidents occurring in your workplace, they can still occur. For example, people can be exposed to chemicals and require immediate medical treatment, a fire can start or a loss of containment can occur. It is therefore necessary to be prepared for any foreseeable incident.

Notes:

Regardless of controls put in place to prevent incidents occurring in your workplace, they can still occur.

译文：无论采取何种控制措施来防止事故在您的工作场所发生，它们仍然可能发生。

语法：句首 regardless of 意为“不管……，不顾……”，put in place 为过去分词短语作 controls 的后置定语。

Reading comprehension

1. What is chemical safety?

2. In order to manage risk, what must a duty holder do?

3. How do we usually identify the identities of chemicals?

Exercise

1. True or false.

（1）Chemicals are widely used and transported in the modern industrial society.

（2）PPE does not reduce risk.

（3）Health monitoring of a person means monitoring the person to identify changes in the person’s health status because of exposure to certain substances.

2. Translate the following sentences into English or Chinese.

（1）The hazardous nature of many chemicals may be increased when mixed with other chemicals, heated or handled inappropriately.____________________

____

（2）应进行的风险评估类型取决于所执行工作的性质。__________

____

Further Reading

Common Safety Practices

1. PPE

Basic chemical safety practice includes wearing protective personal equipment such as safety goggles. Personal protective equipment is a combination of safe work practices, which help to provide sufficient protection from the risks posed by hazardous chemicals, but it is an effective approach to minimize the risk of exposure in controlled environments. Safety goggles are required when handling chemicals to prevent chemicals from getting into eyes. Wearing standard gloves, closed-toed shoes, long trousers, and laboratory coats to protect the stomach, back and forearm is usually required in the laboratory with similar provisions for the workplace. Common chemical PPE are shown in Figure 13-10.

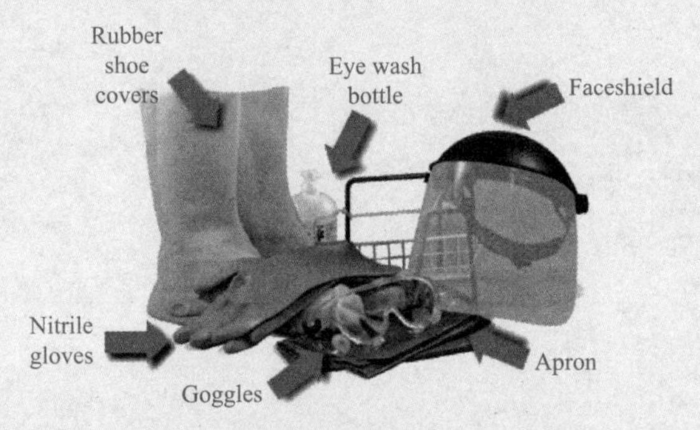

Figure 13-10　Common chemical PPE

2. Labelling

For most of the world, a standard set of illustrative pictograms has been adopted to indicate where hazards exist and the type of hazard present. These pictograms are routinely displayed on containers, transport vehicles, safety advice and anywhere where the material occurs. These have been extended and standardised as the Globally Harmonized System of Classification and Labelling of Chemicals and are now used throughout much of the world. Figure 13-11 illustrates an example of a bottle label.

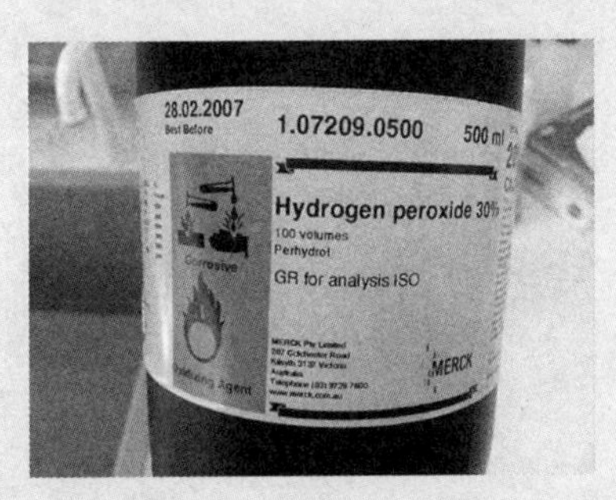

Figure 13-11　An example of a bottle label

Reading comprehension

1. What is PPE? List two or three.

2. What is the scope of application of the Globally Harmonized System of Classification and Labelling of Chemicals?

项目六
专业英语应用与实践

Part Ⅵ:
Application and Practice of CEPE

第十四单元　英文科技应用文
Unit 14　English Scientific Treatises

In this lesson you will learn:

- Features of patents
- Features of instructions
- Features of industrial standards

伴随着科学技术的飞速发展和国际间交流的深化，英文科技应用文发挥着越来越重要的作用。科技应用文是指在科学研究和科技管理工作中所形成的各种应用文的总称，如展示科学研究成果的学术论文；申报科研项目时的立项申请书；项目研究过程中的进度报告书；科学研究成果的鉴定书；维护知识产权的专利申请书；需要在全国某个行业范围内统一的技术要求所制定的行业标准以及对某事或物进行相对的详细描述的说明书等。具体地说是根据国家一定时期的路线、方针、任务和有关科学技术政策、法律、法规，以科学技术为表达对象，以书面语言（包括图、表、公式、数据、符号等）为表达手段，对科学技术领域的各种现象、活动及成果，进行记录、总结、描述、储存、交流、传播和普及，及时沟通科技信息，处理科技领域的各种事务，以推动科学技术进步和国民经济全面、持续、健康发展的创造性的认识和书写的实践活动。

科技应用文种类繁多，本节主要介绍英文专利、英文说明书和英文行业标准的特征及应用。

一、英文专利

专利（patent），字面上的意思是指专有的权利和利益。“专利”一词来源于拉丁语 litterae patentes，意为公开的信件或公共文献，是中世纪的君主用来颁布某种特权的证明。在现代，专利一般是由政府机关或者代表若干国家的区域性组织根据申请而颁发的一种文件。这种文件记载了发明创造的内容，并且在一定时期内产生这样一种法律状态，即获得专利的发明创造在一般情况下他人只有经专利权人许可才能予以实施。

专利文献作为技术信息最有效的载体，囊括了全球 90% 以上的最新技术情报，相比一般技术刊物所提供的信息，专利更具有新颖、实用的特征。科研工作中经常查阅专利文献，不仅可以提高科研项目的研究起点和水平，而且还可以节约 60% 左右的研究时间和 40% 左右的研究经费。

1. 结构特征

各国都有对专利申请文件内容和格式的规定。虽然不同国家的规定有差异，但随着专利制度日益国际化，各国专利文献的格式也日趋相同。以美国的专利申请文件为例，专利申请文件通常包括以下几方面主要内容：

- Title of the invention（发明标题）
- Statement regarding federally sponsored research or development（有关联邦资助的研究或发展的说明）
- The names of the parties to a joint research agreement（各方的联合研究协议的名称）
- Cross reference to related application（交叉参考相关申请）
- Background of the invention（发明背景）
- Brief summary of the invention（发明的简要内容）
- Brief description of the several views of the drawing（对于插图的简要描述）
- Detailed description of the invention（发明的具体描述）
- Claim or claims（权力要求书）
- Abstracts of the disclosure（公开的摘要）
- Drawings（插图）

2. 语言特征

专利文献既是技术文件又是法律文件，它有一些惯用的语言特征，以下做简要介绍。

词汇方面，英文专利常出现惯用词组，例如：Appl. No.（Application Number）申请号；Ser. No.(Series Number) 申请书登记号；Int. Cl.3（International Patent Classification, 3 rd Edition）国际专利分类表（第三版）；Abstract of the disclosure 发明摘要；Preferred embodiments 最佳实施方案；Sheet 3 of 4 第 3 页（图共 4 页）。此外，专利申请文件的起草者会尽可能达到防止被侵权的目标，会极力做到准确、清楚地描述技术方案。如果他们认为现有的名词可能与其本意不完全一致时，会在专利说明书中对这些词的含义进行限定。在专利申请文件的翻译中，时常会遇到专利申请文件起草者对词的重新定义，翻译时要特别注意对这种情况的处理，在充分理解原意的基础上用恰当的手段和适当的汉语表达出来。

句式方面，在英文专利文献中，权利要求部分通常都使用长句。目前各国都会要求一个权利要求是一句话，即只能有一个句号。如果在一项权利要求中有两个以上的句号，就意味着有两个以上权利要求。翻译时不必完全拘泥于英语的句式，可以按照汉语语言习惯，适当变化句子的结构，以求表述易于阅读、更加清楚。

例如：Referring to FIG.1, two-stroke cylinder international combustion engine 10 receives filtered intake air through intake air passage 24 and across intake air valves 16 of butterfly or rotary type and into intake manifold 12 for distribution to engine cylinder（not shown）. 按字面直接翻译："参见图 1，两冲程多缸内燃机 10 接收流经进气通道 24 和蝶形或旋动形进气阀 16 进入到进气歧管 12 分配给各发动机气缸（图中未标示）的经过过滤的进气"。由于原文句子较长，按字面直接翻译译成了有 60 个字的复杂长句，不便于理解。如对句子进行适当拆分和调整，变成若干个短句，则更符合中文的阅读习惯，即"参见图 1，在两冲程多缸内燃机 10 中，过滤后的空气经过进气通道 24 和蝶形（或旋动形）进气阀 16 进入进气歧管 12，分配给发动机的各个气缸（图中未标示）"。

语态方面，在英文专利文献中，为了显示客观公正而大量使用被动句，但是"be ＋过去分词"并不一定都是被动语态，有时是系表结构，表示状态。例如：This substance is known for its hardness. 这种物质以其硬度而知名。

时态方面，与其他科技应用文相似，专利文献中最常用的时态是一般现在时，给人以"无时间性"的概念，以排除任何与时间关联的误解。此外，也会出现一般过去时以及少量的现在完成时和一般将来时。

二、英文说明书

说明书是以应用文体的方式对某事或物进行相对的详细描述，方便人们认识和了解某事或物。说明书要实事求是，不可为达到某种目的而夸大产品作用和性能。说明书要全面地说明事物，不仅介绍其优点，同时还要清楚地说明应注意的事项和可能产生的问题等。说明书可根据情况需要，使用文字、图片、图表等多样的形式，以达到最好的说明效果。按所要说明的事物，说明书一般分为产品说明书、使用说明书、安装说明书、戏剧演出说明书等。

说明书的英译一般都具有准确性（accuracy)、简明性（conciseness）和客观性（objectivity）的共同体现，以下从词汇、句式等方面简要介绍。

1. 词汇方面

一些专业术语、固定用语和习惯用语必须表达得准确、地道，且常使用缩略式简化以避免重复，例如：

中央处理器：Central Processing Unit 缩写为 CPU

液晶显示：Liquid Crystal Display 缩写为 LCD

说明书中还广泛使用复合名词结构。在译文中复合名词结构代替各式后置定语，以求行文简洁、明了、客观，如：

设备清单：equipment check list（不用 the list of equipment check）

保修卡：warranty card（不用 the card of warranty）

英文说明书常使用非人称名词化结构作主语，使句意更客观、简洁，如：

由于使用了计算机，数据计算方面的问题得到了解决。

The use of computers has solved the problems in the area of calculating.

2. 语法句式方面

与其他英文科技应用文相似，说明书的主体部分通常为“无时间性”（Timeless）的一般叙述，因此译文普遍使用一般现在时和被动语态，以体现内容的客观性和形式的简明性，并使读者的注意力集中在受动者这一主要信息上。例如：

本传真机与数码电话系统不兼容。

This facsimile machine is not compatible with digital telephone systems.

您可以在光盘中的电子使用手册中找到额外的信息。

Additional information can be found in the electronic user’s manual which is located on the CD-ROM.

此外，说明书在句式方面一大特点是广泛使用祈使句。电器、电子产品说明书很多地方都是指导用户要做什么、不要做什么或如何做，所以其译文经常使用祈使句，谓语一般用动词原形，没有主语，表述力求准确客观、简洁明了。例如：

请勿将 CF 卡存放在过热、多灰尘或潮湿的环境中，也不能存放在能产生静电荷或者电磁波的环境中。

Do not store CF cards in hot, dusty or humid places. Also avoid places prone to generate static charge or an electromagnetic field.

3. 说明书常用句型

（1）（情态动词）+ be + 容词（或过去分词）+ 的状语，常用于文章开头，说明该产品是做什么用的。例如：

This product can be used in hot water or steam line with the temperature limited to 225℃.

本产品可在热水或温度低于 225℃的蒸气条件下使用。

（2）（情态动词）+ be + 介词短语，用于说明物体的特征、状态和范围，以及计量单位等。例如：

The motor shaft and main shaft should be in correct alignment so as to avoid vibration and hot bearings.

请注意电动机与主轴同心度，以免产生振动或引起轴承发热等情况。

（3）be + 形容词 + 介词短语

Motor pulley is provided with taper sleeve so as to be easy in installing and dismounting it.

电机皮带轮上备有锥套，便于安装和拆卸。

（4）分词和动名词结构，用于说明维修或操作程序及说明有关技术要求。

When operating, don't put your foot on the pedal switch board constantly, so as not to accidentally stop on the switch, causing an accident.

工作时请注意不要经常把脚放在踏板上，以免不慎踏动，引起事故。

（5）名词＋过去分词（或形容词）

the stem bent 阀杆弯曲；the spring broken 弹簧损坏

三、英文标准

标准是由一个公认的机构制定和批准的文件。它对活动或活动的结果制定了规则、导则或特殊值，供共同和反复使用，以实现在预定领域内最佳秩序的效果。制定、发布及实施标准的过程，称为标准化。

标准的制定和类型按使用范围划分为国际标准、区域标准、国家标准、专业标准、地方标准、企业标准，国际标准由国际标准化组织（ISO）理事会审查，ISO 理事会接纳国际标准并由中央秘书处颁布；国家标准在中国由国务院标准化行政主管部门制定。本节以 ISO 国际标准为例，介绍英文标准的结构和常用词汇。

1. 标准的结构

一份国际标准文本中一般包含如下元素：

（1）标题（title，必要元素）

一项国际标准的标题需清楚、简洁地反映出技术提案的范围，清楚地说明标准的主题；标题中最多包含 3 个元素（介绍性元素、主要元素、补充性元素）

例如：Information technology（概括性领域，属介绍性元素）—Automatic identification and data capture techniques（具体领域，属主要元素）—Part 16: Crypto suite ECDSA-ECDH security services for air interface communications（具体的、详细说明，属补充性元素）

译文：ISO/IEC 29167-16:2015 信息技术 自动识别与数据采集技术 第 16 部分：用于空中接口通信的密码套件 ECDSA-ECDH 安全服务

（2）目录（contents，必要元素）

（3）前言（foreword，必要元素）

一般由两大部分组成：通用的内容（general）+ 具体的内容（specific，即简要介绍本标准内容，列出本标准的主要不同之处）。

（4）引言（introduction，可选元素）

该部分提供了标准的背景信息，说明了该标准存在的理由，需保持简短，不要与范围（Scope）的内容重复。

国际标准组织 ISO:

ISO is an independent, non-governmental international organization with a membership of 165 national standards bodies. Through its members, it brings together experts to share knowledge and develop voluntary, consensus-based, market relevant International Standards that support innovation and provide solutions to global challenges.

Find out more: http://www.iso.org

（5）范围（scope，必要元素）

清晰、简洁地告知用户该标准是做什么的，仅描述事实，无要求、建议和许可等相关内容。通常以 This document 作为主语，例如：

This document specifies minimum requirements and test methods for rice (*Oryza sativa* L.).

It is applicable to husked rice, husked parboiled rice, milled rice and milled parboiled rice, suitable for human consumption, directly or after reconditioning.

It is not applicable to cooked rice products.

本标准规定了水稻的最低要求和检测方法。适用于糙米、半熟糙米、精米和半熟精米，适合直接或经过再处理后食用。不适用于熟米制品。

（6）规范性引用（normative reference，必要元素）

在标准或法规中引用一个或多个标准，以代替其具体内容，从而简化标准的编写工作，减少差错或不一致处。通常仅引用公开可用的文档，正常情况下引用的是 ISO、IEC 或者 ISO/IEC 标准。如果该标准文本没有引用标准，此部分描述为“There are no normative references in this document”。

（7）术语和定义（terms and definitions，必要元素）

对文档中使用的、无法自解释且会有不同诠释的术语进行定义，不要定义商标名、古老的术语、口语化术语等。在进行定义时，不需要添加任何冠词，尽量用一句话进行解释。例如：

Disaster：Event that causes great damage or loss.

（8）符号（symbols，可选元素）

在标准文本中，需对文中使用到的所有符号进行定义。按照字母表顺序罗列所有符号，不需要任何介绍性语句。例如：

t_b　　beginning of time interval

t_e　　end of time interval

t_{int}　　time interval

（9）技术内容（technical content，必要元素）

（10）附录（annex，可选元素）

分为规范性附录（normative reference，对正文要求的补充）和资料性附录（informative reference，对正文信息的补充）两种。必须注明该附录属于哪一种，并添加附录标题。即使只有一个附录，也需标记为 Annex A。

（11）参考文献（bibliography，可选元素）

该部分罗列出了标准文本中使用到的提供补充或背景信息的文档。一般罗列参考文献的顺序是：ISO 标准（或其他国际标准）、区域性标准、国家标准、参考文献书目等。

示例 1　参考文献为标准：标准类型 标准号，年份（如需要），标题—分标题 ISO/IEC 29167-16, 2015, Information technology—Automatic identification and data capture techniques—Part 16: Crypto suite ECDSA-ECDH security services for air interface communications

示例 2　参考文献为期刊杂志：作者姓（全大写）名缩写（首字母大写）.

文章标题 . 杂志名称 . 版权年，卷号 PP. 起页码-止页码

GRUN E., ZOOK H.A., FECHTIG H., *et al*. Collisional Balance of Meteoritic Complex. Icarus. 1985, 62 pp. 244-272

示例 3　参考文献为书籍：作者姓（首字母大写），名缩写大写 . 标题 . 出版地 . 出版单位，出版年

Allen, T.B. Vanishing Wildlife of North America. Washington, D.C. National Geographic Society, 1974

2. 英文标准常见用语

（1）前言部分

本国家标准等同采用 IEC（ISO）×××× 标准 :

This national standard is identical to IEC（ISO）××××

本国家标准修改（等效）采用 IEC（ISO）×××× 标准 :

This national standard is modified in relation to IEC（ISO）××××

本国家标准非等效采用 IEC（ISO）×××× 标准：

This national standard is not equivalent to IEC（ISO）××××

本国家标准对先前版本技术内容作了下述重要修改：

There have been some significant changes in this nationals standard over its previous edition in the following technical aspects

本国家标准与所采用国际标准的主要技术差异：

The main technical differences between the national standard and the international standard adopted

本国家标准从实施日期起代替 ×××× ：

This national standard will replace ×××× from the implementation date of this standard

（2）引言部分

主题和范围

subject and the aspect(s) covered

本国家标准规定……的尺寸

This national standard specifies the dimensions of…

本国家标准规定……的方法

This national standard specifies a method of…

本国家标准规定……的性能

This national standard specifies the characteristics of…

本国家标准规定……的系统

This national standard establishes a system for…

本国家标准规定……的基本原理

This national standard establishes general principles for…

本国家标准适用于……

This national standard is applicable to…

第十五单元　求职申请与简历撰写

Unit 15　Preparation of Cover Letter and Resume(CV)

In this lesson you will learn:

- Features of cover letter
- Features of resume

求职是大学生走出校园、迈向社会的关键一步。在求职过程中，求职申请和简历能够让招聘者在短时间内发现求职者的优势和亮点，给用人单位留下良好的第一印象。本节将结合范例介绍英文求职申请和简历的撰写。

一、求职申请撰写

在向外资企业、外国组织或机构求职时，求职者应随简历附上一封内容精练、有针对性的求职申请（cover letter）。英文求职申请的撰写需要写作者仔细阅读招聘启事中对岗位的描述和对求职者的要求，了解岗位所需的能力和素质，在写作时有的放矢地展现出自己具有用人单位需要的知识、能力、经历和业绩。

1. 求职申请的基本格式和内容

求职申请的长度一般不超过一页，格式与英文正式信函大致相同，包括求职者姓名和联系方式、招聘方姓名和联系方式、称呼、正文、结束语和签名。基本格式如下：

YOUR NAME

[Your phone number] [Your email address] [Your mailing address]

[Hiring manager's name]

[Company address]

[Company phone number]

[Hiring manager's email address]

Dear [Mr./Ms.] [Hiring manager's name],

Opening paragraph that mentions how you found out about the job and why you're interested in the position. Make sure you use the correct job title here, or else the employer will think you're sending out a generic cover letter.

Body paragraph(s) that provide examples of your professional accomplishments, skills, and work experience. These examples should all tie into why you're the best fit for the role.

Closing paragraph that expresses your interest (again), and restates why you're the right candidate for the job. You can mention whether you prefer to be contacted by phone or email here as well. If you prefer phone calls, provide the times when you're available.

[Sincerely /Best regards],

[Your signature]

需要特别注意的是，在撰写英文求职申请时，求职者应仔细查看招聘信息，把收信人的名字写得具体准确，尽量不用 Sir/Madam 等统称，否则会给人留下一信多投的不良印象。求职申请的正文可分为三部分：首段说明自己应聘的职位，表明自己如何得到招聘信息并表达求职意愿。主体段展示自己的职业和学业成就、技能和经验等，写明与岗位相关的最令人印象深刻的素质和能力。结尾段强调自己是适合岗位的人选，重申求职意愿，明确表达对面试的期待，还可以给出联系面试的方式和时间等信息。

求职申请首段常用的句式包括：

- I am writing to apply for the position of ...
- I learn from your advertisement in the newspaper that... is in need of a(n)...
- The position of ... that you posted on the Internet drew my attention...
- Please accept this letter and resume as my application for the position of...
- As a long-time follower of your company, I was excited to see the opening for the... position on your website.
- I believe that my education and recent job experience make me the ideal candidate for the position.

求职申请主体段常用的句式包括 :

- As you can/will notice in my attached resume, ...
- My professional experience has given me the opportunity/ skills/ ability...
- I know your company is looking for..., and I have the experience/track record of...
- My experience and education provide me with excellent knowledge of ... and skills of ... required for this position.
- I have won awards for excellence in ... and feel that I can utilize these same skills in your company.

求职申请结尾段常用的句式包括：

- Thank you for your time and consideration of my qualifications.

- I look forward to hearing more about the position and appreciate your consideration.
- My resume is enclosed for your review and consideration.
- I believe I am the best and most qualified candidate for the job.
- Please contact me at your earliest convenience to arrange an interview to further discuss my qualification.
- I would appreciate the opportunity for an interview with you and thank you in advance for your time and consideration.

2. 求职申请范例

Zhao Chang

（123）456-7891 zhaochang@example.com Tianjin, China, 300000

Mr. Aiken

Jerome Laboratories

(000)123456

aiken007@example.com

Dear Mr. Aiken,

I am writing to apply for the Assistant Chemist position with Jerome Laboratories, which I saw advertised on *ChemistryJobs.com*. I hope to get an opportunity to work in your lab and I feel that my education and recent job experience make me the ideal candidate for the position.

I graduated from Tianjin Vocational Institute with 3 years in the chemical industry working directly in the lab. In my current position as Assistant Chemist with Standard Chemicals I help conduct experiments, maintain laboratory safety and document findings. I am thorough in my record keeping and pay great attention to detail traits that are essential in this field. In addition I always adhere to safe handling procedures and maintain sanitation protocol. My expert knowledge of chemistry is what sets me apart from the competition. As you can see in my resume, I have earned accolades and awards for my accomplishments on the job as well as several academic honors in the field of chemistry. I would love the chance to bring my expertise to Jerome Laboratories as your new Assistant Chemist.

I believe I am the best and most qualified candidate for the job. Please contact me at your earliest convenience to arrange an interview to further discuss my qualifications. I am anxious to speak with you and thank you in advance for your time and consideration.

Sincerely,

Zhao Chang

sanitation *n.* 卫生

protocol *n.* 方案，协议

accolade *n.* 荣誉

expertise *n.* 专业知识，专业技能

M15-1
单词读音及例句

二、简历撰写

求职时，好的简历往往是通往成功的大门，一份专业准确、重点突出的英文简历会给雇主留下良好的第一印象。简历的本质是简洁的个人描述性文书，因此好的简历应该简洁清晰、主次分明，具有针对性，让招聘者对求职者的经历和技能一目了然，快速判断出求职者是否适合岗位。通常来说，招聘者在每一份简历上所花的时间很短，一般不超过30秒，有时甚至只有8至15秒的浏览时间。因此，用简历在短时间内体现自己的优势和专业程度对求职者至关重要。

1. 简历的基本格式和内容

英文简历是对求职者个人基本信息、求职意向、教育背景、工作经历、资格证书、成就特长等方面的简要介绍，要求内容客观准确，文字简练，版式清晰，方便阅读。简历的长度一般是一页纸，最长不超过两页纸，格式没有一定之规。但需要特别注意的是，简历制作的主要原则是方便招聘者阅读，因此主要内容的标题应适当放大加粗，突出重点，文字排列工整清晰，不要过度追求美观的设计，让阅读者眼花缭乱。

简历的内容主要包含个人基本信息、求职意向、教育背景、工作经历和附加信息这五部分：

① 个人基本信息部分包括求职者的姓名、电话、住址、电子邮箱等，注意英文地址小单位在前大单位在后，邮编放在省市名之后。电话前面要加地区号，区号可用括号或与号码间加空格。如：（8610）650-5226；较长的号码中间可以加“-”分隔开，如6505-2266，138-0135-1234。

② 求职意向部分无须长篇大论，用短句或短语有针对性地写下相关领域或具体岗位即可：如 Primary school science teacher; To obtain a secretarial position; Customer service representative in health-care industry 等。

③ 教育背景部分主要包含以下信息：学校名称（大写加粗）及所在的城市、州 / 省、国家、（预计）毕业年月、所获文凭的类型和级别，根据个人情况也可以列出 GPA（grade point average）、相关的专业课程、在校期间所获得的奖学金、奖项等。教育背景通常从大学写起，个别从优秀高中毕业且缺乏工作经验的求职者可以从高中写起，按照时间倒序排列，最近的学历放在最前面。

④ 工作经历部分是简历中最重要的部分，因为招聘者往往对求职者的职场经历和价值更为看重。工作经历中所涵盖的内容主要是实习或者全职工作的经历，不过，岗位相关的志愿者经历和校内职业导向的社团经历也可以算作是工作经历的一部分，作为实习经历不足情况下对职业能力的补充展示。这部分应写明实习 / 工作单位及所在地、实习 / 工作起止时间、实习 / 工作职位名称及性质、实习 / 工作内容及相关成果，多条工作经历按照时间倒序排列，最近的经历放在最前面。工作单位名称用黑体加粗表示，选用单位的官方译名，如果没有官方译名可按照公司的业务类型进行

翻译，最好不要用拼音翻译单位的名字。在描述工作内容和成果时，少用长句，用简洁的语言和具体事例、数字来证明自己的工作业绩，如：Serving customers at the sales counter; Providing face-to-face advice to customers on the store's products; Exceeding the required sales goal by 30%。

⑤ 附加信息是对个人特点和资质的进一步展示，可根据个人情况写下技能、爱好特长、个人荣誉等信息，如技能中可列出所获的职业资格证书、语言等级证书、计算机等级证书等。需要注意的是，简历具有很强的针对性，因此不要罗列太多与应聘岗位无关的信息。

2. 简历范例

FRANK WU

Personal Info: 16 Jinhai Street, Tianjin 300000 • 123-456-7890 • frankwu@gmail.com

Objective: Assistant Chemist

Work experience

April 2018 - July 2019　Chemistry Laboratory Assistant

Jerome Laboratories, Tianjin

- Assisted 2 professors in experiments pertaining to Organic structure and reactivity subjects.
- Tested, documented and analyzed all experiments with a 99.9% error-free documentation record.
- Prepared and stored 25+ reagents to complete different experiments every week.

2010.7 - 2015.9　Chemistry Laboratory Assistant

Growthsi Lab, Tianjin

- Implemented the use of MS Excel pivot tables to identify laboratory supplies, saving ¥30000 annually in supply costs.
- Analyzed and recorded data for 75 samples on a weekly basis.
- Handled the inventory for all laboratory supplies and chemicals.

2007.9 - 2010.6　Volunteer Laboratory Assistant

Middle West Laboratories, Tianjin

- Organized and synchronized a team of volunteers.
- Helped implement a new system for storing laboratory test data, saving 3 hours per week in documentation.
- Ordered facilities with 100% accuracy and on time.

Education

2003.9 - 2007.7　Bachelor of Science

Tianjin University

Major in Applied Chemistry

Skills

Fluent in English and French　　Expert in Microsoft Office

CET-6 certificate

Honors and Awards

- First Prize, Personal Achievement Award, Tianjin Volunteering Organization, 2017
- Outstanding Graduate, Tianjin University, 2007

第十六单元　英文面试技巧

Unit 16　Interview Skills

In this lesson you will learn:

- Interview skill
- Expressions used in interview

面试是求职过程中的关键环节。近年来，许多用人单位要求求职者参加英文面试，而不少求职者凭着良好的教育背景和丰富的工作经验一路过关斩将，却往往在英语面试的环节上功亏一篑，最终无法实现自己的职业理想。本节将介绍英文面试的技巧和常用表达，并为化工专业英语学习者提供化工行业英文面试中的一些常见问答。

一、面试技巧

1. 技巧一：在面试前做好充分准备

英文面试中求职者的心态调整至关重要，因为目前环境下绝大多数参加英文面试的求职者都处于有一定交流基础但经验不足的水平，面试官考察的是求职者的综合素质，而不仅仅是语言能力，对于求职者的英文口语水平没有过高的要求，所以求职者只要展现出沉稳自信的态度、条理清晰的回答和满足基本交流的口语水平就可以给面试官留下足够良好的印象。

要做到谈吐镇定自若，回答有条有理，需要求职者在面试前做好充分的准备，多做模拟练习，不要抱侥幸心理，在面试时全靠临场发挥容易因过度紧张而导致失控，让自己的表现大打折扣，无法发挥真实水平。在做英文面试准备时不能只专注自己的发言，还要注意训练听力，至少要听得懂面试官的问题。与此同时，求职者应避免死记硬背，在英文面试中大段背诵准备好的材料，让回答流于死板生硬，也不要舍本逐末，在面试表达中加入大量长难句展示语言能力。在面试中，简单、直白、清晰的回答往往最能高效率地展现求职者的能力和素质。

2. 技巧二：注意英语时态的变化

求职者在参加英文面试前大都作过充分的语言知识准备与练习。在众多的英语语法规则中，之所以要单独强调时态的运用，是因为和汉语的表达习惯不同（汉语中动词没有时态变化），时态错误是英文口语表达中最常见的错误之一。同时时态又是英语比较基本的语法点，一旦用错，会让面试官对面试者的英语能力产生质疑。除此之外，在面试过程中，往往会涉及很多关于个人经历、教育背景、工作经验、职业规划等方面的问题，因此在表述某件事情或是某个想法的时候，一定要注意配合正确的时态，否则就会造成“差之毫厘，失之千里”的后果。例如，已经参加过某项专业技能培训与正在参加或计划参加在英文中的时态表达完全不同。

3. 技巧三：以英语为载体，展示工作才能

一般而言，对于非英语专业要求的工作，英文面试与英语考试中的口语考试不同，面试人员通常是由公司的人事主管、应聘部门主管或公司高层组成，他们更关心和器重的是你的专业知识和工作能力，而英语此时只是一种交流工具，或者说是你要展示的众多技能中的一种。因此要切忌为说英语而说英语，有些求职者怕自己的英语减分，希望给面试官留下英语水平高的印象，常常会大量使用事先准备好的花哨的词汇及句式，而真正针对面试官所提问题的、与工作有关的个人见解却很少，内容空泛，逻辑混乱，言之无物。求职者在面试中无须夸夸其谈，只需用简单清楚的英文说明自己的教育背景和经历，介绍自己之前的成就，充分展示自己的热情和能量，同时展现自己对岗位的兴趣，强调可以为公司带来的价值。

二、英文面试常用表达

1. 英文面试常见问答

（1）What can you tell me about yourself?

I recently graduated from South University, where I led 150 other peers as the class president, and implemented the campus's first recycling program. Last summer I interned at San City Corporation, where I developed the holiday sales and marketing strategy that helped lead the company to the largest growth in sales they had ever experienced. I really enjoyed that experience. In my senior year I also participated in an internship at Middle West Lab and stayed there for 3 months. I believe I have the knowledge and skills needed to do well in the field. I hope I'll have a chance to get to know you and work with you in the future.

（2）Why did you apply for this job?

I decided to apply because I really like the vision of your company and the career growth possibilities. I can use my full potential at this position and help your company to grow and prosper.

（3）Can you tell me something about your education?

I have studied at ABC University. I was active during my studies and took part in practical projects and courses too. Overall my studies were pretty practical and I believe to be ready for job of a project manager.

（4）Why should we hire you?

I have relevant working experience and I am strongly motivated to work for your company, as it has always been my dream to work here.

（5）What are your strengths?

I am very responsible and always accomplish all my duties on time.

I am pretty organized, which is reflected in the quality of my work.

I have good communication skills. I believe that communication skills are crucial in every job, but especially in this one.

（6）What are your weaknesses?

I trust people too much. It is nice to live with such feelings, but it caused me many troubles in the past. However, as I am getting older, I can distinguish friends from foes much better .

（7）How would your friends or colleagues describe you?

They say Ms. Wang is an honest, hardworking and responsible woman who deeply cares for her family and friends.

（8）What are your goals in five years horizon?

To be honest, I'm very interested in leadership opportunities. 5 years down the line, I would love to be in a manager's position, training up junior staff and helping them develop and make a positive contribution to the company.

My goal is to become a better manager and to help my employer to achieve exceptional results as a company.

I want to start a family and have a good job. That's all I want. I believe that your company is a right place to start realizing my goals.

（9）What are your salary expectations?

Money is important, but the responsibility that goes along with this job is what interests me the most.

To be frank and open with you, I like this job very much, but I have a family to support, so I will accept an average salary for this position which is something between $35,000 and $40,000.

（10）Do you have any questions?

Do you have a training program for new employees?

Can you tell me something more about the working environment?

What are the company goals and how can I look to help achieve these goals?

2. 化工行业英文面试常见问答

（1）Why did you apply for this job as a chemist?

Throughout my education, I have always been drawn to the experimentation aspects of chemistry. I love the hands-on problem-solving approach, along with the opportunity to use new and varying materials to develop innovative solutions. Before graduating college, I participated in several internships working within the pharmaceuticals industry.

Through these internships, I gained significant experience collaborating with other chemists and adhering to good manufacturing practices to ensure high-quality work. I want to work in this industry because it allows me to use my passion for chemistry to help others. Your company represents one of the leaders in this industry, and I would love the opportunity to work alongside your team of experts and use my skills and knowledge to support them.

（2）What skills or characteristics make you qualified for a chemist role at our organization?

As someone who has worked within the chemistry field for several years, I understand the importance of being safety conscious. I have strong attention to detail, which I use to maintain a safe environment for myself and my colleagues. In fact, I have spent time training less experienced colleagues on several safety rules, such as proper glassware-cleaning techniques. I also use this skill when recording notes and documentation. My prior supervisor complimented me on my work regularly due to its detail and accuracy.

（3）How do you manage conflicts with your colleagues in the lab?

If I encounter a problem with a colleague, I initially try to resolve it with him/her privately. A couple of years ago, I had a colleague who always left a mess in the lab. I pulled the person aside and explained that his/her behavior negatively affected the work of myself and our other colleagues. It disrupted our processes because we had to spend time cleaning up the mess before starting our work. I also explained that the mess could also lead to safety issues in the lab.

Luckily, my colleague heard and understood my concerns and promised to be more careful in the future. Setting those expectations helped motivate him to follow our lab etiquette rules more carefully. However, I realize that conflicts cannot always be resolved so easily. If this colleague had not changed his behaviors after our conversation and several reminders, I would have taken my concerns to a supervisor to gain his/her help in resolving the issue.

（4）How do you stay up-to-date on news and innovations within the chemistry field?

I am passionate about chemistry, and I use my free time to learn about the latest happenings in our industry. In the morning, I often browse some of my favorite scientific journalism websites. When I read something interesting, I enjoy bringing up the topic at work and discussing it with my colleagues. I've found that sometimes what we read ends up benefiting our work. I also follow several chemistry thought leaders on social media and participate in online conversations about these topics.

（5）Are you familiar with SOP and GMP?

SOP stand for Standard Operating Procedures, and GMP stands for Good Manufacturing Practice. SOP defines the safe handling of chemicals within a lab, while GMP represent the methods and processes for producing goods consistently according to specific quality standards. In my current position as an associate chemist, I hold responsibility for developing and implementing SOP in our lab. I work in pharmaceuticals, so I also use GMP to conduct and document our team’s work to ensure its quality.

附录
《化工专业英语》(Part Ⅱ -Part Ⅴ) 习题答案

Part Ⅱ -Unit 5-Lesson One Answers

Reading comprehension

1. The atomic number is associated with an element's chemical behavior.

2. Two.

3. The periodic table arranges the elements with similar characteristics in the same vertical column.

4. He, Ne, Ar, Kr, Xe, Rn, Og.

Exercise

1.（1）True.（2）True.（3）True.

2.（1）通过这种排列方式，同一族（竖排）的元素具有相似的化学性质。

（2）Most elements, including all transition elements, are metals.

Further Reading

Reading comprehension

1. +2.

2. These elements form such a variety of different compounds (exhibit different chemical valencies).

3. The elements of group ⅦA.

Part Ⅱ -Unit 5-Lesson Two Answers

Reading comprehension

1. Ammonium nitrate. It is used as a soil fertilizer.

2. Most organic compounds is comprised of carbon chains that vary in length and shape.

3. Cells use nucleic acids to code the genetic information of an organism.

Exercise

1.（1）False.（2）True.

2.（1）通常，正负电荷相抵形成中性化合物。

（2）Most organic compounds is comprised of carbon chains that vary in length and shape.

Further Reading

Reading comprehension

1. A VOC is a chemical or compound that contains such vapor pressure that

it does not take much heat to vaporize the particular chemical or compound into a gaseous form. For example, methane.

2. VOC mainly affect human respiratory system and immune system.

Part Ⅱ -Unit 5-Lesson Three Answers

Reading comprehension

1. The Brønsted theory defines an acid as any species that can donate a proton; and a base as any species that can accept a proton.

2. Because the numbers of $[H^+]$ and $[OH^-]$ are usually large and inconvenient to deal with.

3. $HPO_4^{2-}(aq)+H_2O(l) \rightleftharpoons H_2PO_4^-(aq)+OH^-(aq)$

Exercise

1.（1）True.（2）False.（3）True.

2.（1）弱酸的强度取决于它解离的程度：解离程度越大，酸性就越强。

（2）Weak acids and bases ionize only partially, and the ionization reaction is reversible.

Further Reading

Reading comprehension

1. Vinegar.

2. The strength of an acid depends on the amount of $[H^+]$, which is determined by both acid strength and concentration.

Part Ⅱ -Unit 6-Lesson One Answers

Reading comprehension

1. The energy absorbed or emitted by a system undergoing some kind of process must be accounted for in all the different kinds of energy, $\Delta U=Q+W$.

2. Total entropy change for a process is taken as the criterion for spontaneity.

3. At absolute zero temperature the absolute entropies of all perfectly crystalline materials are identical and arbitrarily set at 0.

Exercise

1.

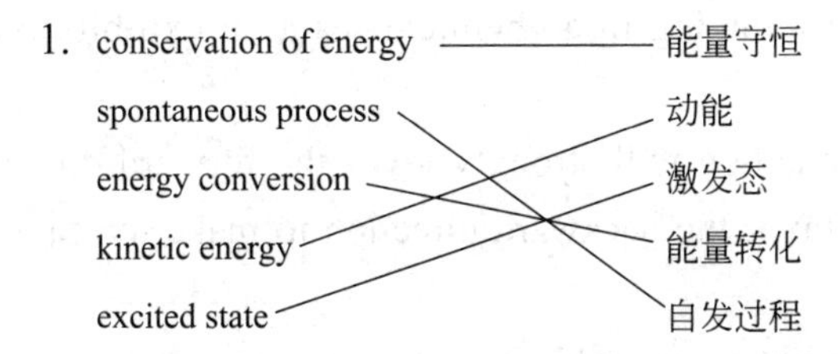

2.（1）热力学第二定律的一种说法是，热量不能自发地从低温物体转移到高温物体。

（2）When heat flows, the temperature of the hotter object decreases and the colder object heats up.

Further Reading

Reading comprehension

1. He used the principle of thermal expension and cold contraction.

2. It appeared as early as the pre-Qin period.

Part Ⅱ-Unit 6-Lesson Two Answers

Reading comprehension

1. A spontaneous process is one that occurs without the addition of external energy.

2. When $\Delta G<0$, the reaction is spontaneous.

3. Ice melts at the temperature of $0^{o}C$.

Exercise

1.(1) False.(2) True.(3) True.

2.(1) 当用吉布斯自由能来判断一个过程的自发性时，我们只关心 G 的变化，而不是它的绝对值。

(2) We can use the sign of ΔG to figure out whether a reaction is spontaneous in the forward direction, backward direction, or if the reaction is at equilibrium.

Further Reading

Reading comprehension

1. It was published in 1876; the title is "On the Equilibrium of Heterogeneous Substances".

2. Henry Adams called Gibbs "the greatest of Americans, judged by his rank in science". Because Gibbs has made great achievements in the science of statistical mechanics, his application of thermodynamic theory converted a large part of physical chemistry from an empirical into a deductive science.

Part Ⅱ-Unit 6-Lesson Three Answers

Reading comprehension

1. Chemical equilibrium is the condition which occurs when the concentrations of reactants and products participating in a chemical reaction exhibit no net change over time.

2. K_c tells us whether a reaction will strongly favor the forward direction to make products or strongly favor the backward direction to make reactants or somewhere in between.

3. When a system at equilibrium is subjected to a change in temperature, volume, concentration, or pressure, the system readjusts to partially counter the effect of the change, resulting in a new equilibrium.

Exercise

1.(1) True.(2) False.(3) True.

2.（1）反应物和产物的量已经达到一个恒定的比例，但它们不一定相等。

（2）Le Châtelier's principle can be used to predict the shift of equilibrium of reversible reactions.

Further Reading

Reading comprehension

1. Decrease.

2. According to Le Châtelier's principle, when a stress (heat in this case) is applied to a system in equilibrium, the equilibrium will shift to the side which is endothermic so as to relieve the stress (heat in this case).

Part Ⅱ-Unit 6-Lesson Four Answers

Reading comprehension

1. Chemical kinetics deal with rates of chemical reactions.

2. The rates of reactions are usually defined as the change of reactant and product concentration per unit time.

3. 3.

Exercise

1.（1）True.（2）True.（3）True.

2.（1）反应速率通常定义为单位时间内反应物和产物浓度的变化。

（2）In a zeroth-order reaction, the rate is independent of the concentration of the reactant.

Further Reading

Reading comprehension

1. Temperature.

2. Commonly used spectroscopic methods include emission spectroscopy, mass spectrometry, gas chromatography and nuclear magnetic resonance.

Part Ⅱ-Unit 6-Lesson Five Answers

Reading comprehension

1. Increase surface area such as crushing a solid into smaller parts.

2. An increase in the concentrations of the reactants will usually result in the corresponding increase in the reaction rate.

3. The catalyst increases the rate of the reaction by providing a new reaction mechanism to occur within a lower activation energy.

Exercise

1.（1）True.（2）True.（3）False.

2.（1）"经验法则"认为温度每升高 10℃，化学反应速率就会增加到原来的两倍，这是一种常见的误解。

（2）The effect of temperature on the reaction rate constant usually obeys the Arrhenius equation.

Further Reading

Reading comprehension

1. Activation energy is the lowest energy needed for the reaction to occur. It can be described as a kind of "hump" you have to get over to get yourself out of bed.

2. The Arrhenius equation relates activation energy to the rate at which a chemical reaction proceeds:

$$k=A\exp[-E_a/(RT)]$$

Part Ⅲ-Unit 7-Lesson One Answers

Reading comprehension

1. The key component in any process is the chemical reactor; if it can handle impure raw materials or not produce impurities in the product, the savings in a process can be far greater than if we simply build better separation units.

2. If it started out spatially uniform, then stirring is not necessary.

3. It might seem that, since the concentration changes instantly at the entrance where mixing occurs, reaction occurs there and nothing else happens in the reactor because nothing is changing.

Exercise

（1）化学反应器是发生化学反应的单元。

（2）只要反应器的体积不变，所写的方程式就是正确的。

Further Reading

Reading comprehension

1. Batch processes.

2. Batch processes can be tailored to produce small amounts of product when needed. Batch processes are also ideal to measure rates and kinetics in order to design continuous processes. Batch process is easier to "tune" slightly to optimize each batch.

Part Ⅲ-Unit 7-Lesson Two Answers

Reading comprehension

1. Tubular reactors are widely used in exothermic gas-phase reactions that require a solid catalyst.

2. A pipe through which flows without dispersion and maintains a constant velocity profile. No mixing because at low flow rates the flow profile will be laminar.

3. Cocurrent flows.

Exercise

1.（1）False.（2）True.（3）True.

2.（1）用于这种反应的设备实际上看起来与换热器相似。

（2）一些较小的反应器的成本往往比较大反应器的成本还高。

Further Reading

Reading comprehension

1. Reactor diameter is calculated from the given pressure drop.

2. The large heat-transfer area associated with small-diameter tubes.

Part Ⅲ-Unit 7-Lesson Three Answers

Reading comprehension

1. For simple reactions, the effect of the presence of a catalyst is to increase the rate of reaction, permit the reaction to occur at a lower temperature, permit the reaction to occur at a more favorable pressure, reduce the reactor volume, and increase the yield of a product(s).

2. The packed bed reactor is the catalytic packed bed reactor. Most of these reactors are designed with an internal lining and a distributor to control the gas passage through the catalyst bed. The fluidized bed is a situation where the fluid and the catalyst are stirred instead of having the catalyst fixed in a bed.

3. As the fluid velocity is increased, the reactor will reach a stage where the force of the fluid on the solids is enough to balance the weight of the solid material. This stage is known as incipient fluidization and occurs at this minimum fluidization velocity.

Exercise

1.（1）True.（2）False.（3）False.

2.（1）生物催化剂毫无疑问是最重要的催化剂（对我们来说），因为没有它们，生命将是不可能的。

（2）Both fixed bed and fluidized bed reactors are commonly used in industry.

Further Reading

Reading comprehension

1. The main advantages of FBMRs include: isothermal operation; negligible pressure drop, and no internal mass and heat transfer limitations; flexibility in membrane heat transfer surface area and arrangement of the membranes; reduced axial gas back-mixing and enhanced bubble breakage.

2. The limitations of FBMRs include: difficulties in reactor construction and membrane sealing at the wall. Erosion of reactor internals and catalyst attrition.

Part Ⅲ-Unit 8-Lesson One Answers

Reading comprehension

1. The spring-tube, spring-plate, and spring-capsule that is slightly deformed by changing pressure.

2. The standard orifice plate, the standard jet, and the standard venturi jet.

3. No.

4. $h = \Delta p/\rho g$

5. Thermocouple thermometer.

Exercise

1.(1)1.013×10^5; 1.(2)Elastic pressure sensor.

2.(1)主要问题是测量相当长时间的超声波，并具有足够的精度。

(2)孔板流量计仅限于测量洁净流体的流量。

(3)When using a thermocouple thermometer, it is necessary to keep the cold junction temperature constant.

Further Reading

Reading comprehension

1. A local device needs to be read at its point of installation, and has no means of signal transmission. And a remote device can automatically collect and record measurement parameters.

2. For example, a flue gas O_2 measurement might output a smoothed version of the actual composition. A brief O_2 deficiency might not be seen, but could be enough to start combustion when the uncombusted vapours meet O_2 elsewhere in the ducting.

Part Ⅲ-Unit 8-Lesson Two Answers

Reading comprehension

1. Control here means that the measured value of a process variable, the so-called control variable x, is maintained at a given setpoint w by adjusting actuators such as valves and pumps by means of a control element until the deviation between x and w is compensated.

2. P, I, and D controllers， and the combinations of three types.

Exercise

1.(1)False.(2)False.(3)False.

2.(1)控制是系统中输入量根据逻辑运算影响输出量的过程。

(2)它考虑了输入的信号状态，并根据计算操作分配输出。

Further Reading

Reading comprehension

1. In the MANUAL mode, it is possible to adjust the control action output directly, using the ‘manual loading station’ at the bottom of the controller. In AUTO mode, the device makes use of its internal computation to set a suitable control action output which will bring the feedback signal to the setpoint value.

2. For the analogue devices, one accessed the back of the cabinet to adjust the proportional gain, integral time constant or derivative time constant. These can of course be set remotely in the computer devices.

Part Ⅲ-Unit 8-Lesson Three Answers

Reading comprehension

1. There will be many manually operated valves on a plant for less frequent use. And the purpose of manual block valves is to prevent residual process fluids from escaping through the vacancy.

2. In control scheme design, it is important to make this specification to ensure as orderly a shutdown as possible in the event of disaster. And the air-to-open (AO) is taken as fail-closed (FC) and the air-to-close (AC) is taken as fail-open (FO).

Exercise

1.

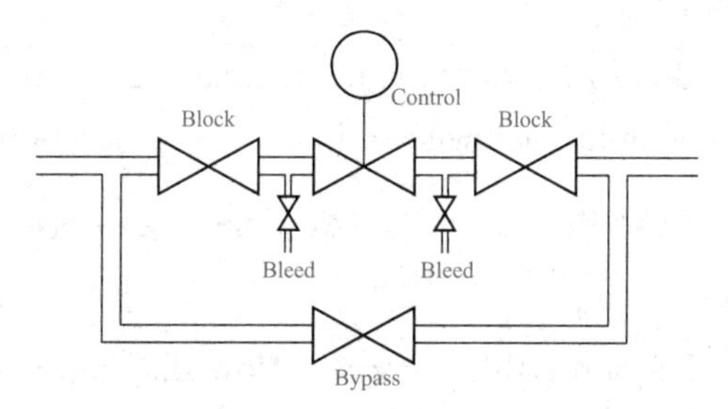

2.(1)The actuator can be operated not only automatically by the control system, but also manually.

(2)从某种意义来说，只需要对气开型执行器稍微改造就能成为气关型执行器。

Further Reading

Reading comprehension

1. They are robust drives whose high power density results in good control of complex motions, accurate positioning under load, and wide speed ranges.

2. Continuous control elements effect a steady or gradual change in mass or energy flow within the design range, while switching elements act in a step-by-step manner.

Part Ⅲ-Unit 8-Lesson Four Answers

Reading comprehension

1. The comparator compares the set-point with the measured signal (or variable).

2. The difference between the set-point and the measured variable is the previously defined error. Based on the magnitude of the error, the controller element in the feedback loop takes corrective action by adjusting the value of a process parameter, known as the manipulated variable. Unlike feedback control, a feedforward control measures the disturbance directly, and takes preemptive action before the disturbance can affect the process.

3. Two control loops.

Exercise

1.（1）True.（2）True.（3）False.

2.（1）前馈控制很可能适用于某一干扰变化频繁的工艺场合。

（2）Although the cascade control structure is more complicated, it could have much better control effects.

Further Reading

Reading comprehension

1. The objective is to take advantage of the plant-wide view of outputs, and access to inputs, of these computers, in order to enhance regulation and optimisation. In this way, industries have been able to work safely with narrower specifications and less loss.

2. The advanced algorithms focus on criteria such as throughput, product specifications and economics, not necessarily smooth process operation.

Part Ⅲ-Unit 9-Lesson One Answers

Reading comprehension

1. Basic flow diagram (BFD); Process flow diagram (PFD); Process and instrumentation diagram (P&ID) .

2. For clarity reasons, merely the main steps, plant lines and complete offsites are depicted as rectangles with plain text identification. The main mass flows are shown as simple lines with flow direction arrows.

Exercise

1.（1）False.（2）True.（3）False.

2.（1）设备的布置反映了流程图中描述的流程逻辑顺序。

（2）管道设计人员使用管道规范按照流程图所示在两个容器之间布置管道。

Further Reading

Reading comprehension

1. No.

2. On a process flow sheet, the equipment is arranged logically to show the flow of materials through the process, but a photograph shows the final physical arrangement determined by structural requirements without regard to the process flow.

Part Ⅲ-Unit 9-Lesson Two Answers

Reading comprehension

1. All essential components such as pumps, compressors, heat exchangers, columns, vessels *etc.*;

The main pipelines ;

Transport facilities ;

Measuring instruments;

Control devices.

2. Flow (F); Level (L); Pressure (P); Temperature (T).

3. LC (liquid level controller).

Exercise

1.（1）The instrument group is L (Level), and the instrument types are A (Alarm) and H (High) .

（2）The instrument group is T (Temperature), and the instrument types are R (Recorder) and C (Controller) .

（3）The instrument group is P (Pressure), and the instrument types are I (Indicator) and C (Controller) .

2.（1）除了设备符号和工艺流线之外，详细的工艺流程图还有几个基本组成部分。

（2）“PD”将指示工艺过程中两点之间压差的测量值。

Further Reading

Reading comprehension

1. No divider; Single solid line; Single dashed line; Double solid line; Double dashed line.

2. Single solid line.

Part Ⅲ-Unit 9-Lesson Three Answers

Reading comprehension

1. It is the abbreviation of piping and instrumentation diagrams.

2. A P&ID set is developed by the engineers during the design step of the project but will later be used in a process plant during operations. A P&ID may need to be read by engineers, managers with management degrees, trade practitioners, and many others. It also should give enough information about the project. A P&ID can be used during normal operations or an emergency event.

3. It will be based on the prevailing line rule (Horizontal process lines > Vertical process lines > Instrument signals).

Exercise

1.（1）True.（2）False.（3）False.（4）True.

2.（1）重要的是要知道 P&ID 将用于工厂操作过程，包括紧急情况处理。

（2）现场的直管段可以表示为 P&ID 上具有多个方向变化的线。

Further Reading

Reading comprehension

1. P&ID development basically means developing the BFD, the PFD, and then the P&ID. As the process flow diagrams (PFD) only contain the main components, fittings, pipes and measuring points, the P&IDs have now to be

completed.

2. Eye washes; Safety showers; Safety valves; Burst disks; Safety - relevant measuring points.

Part Ⅳ-Unit 10-Lesson One Answers

Reading comprehension

1. Density and viscosity are the most important physical and transport properties in fluid flow studies.

2. No. The Bernoulli equation is suitable for an incompressible inviscid fluid, whereas gases are compressible.

3. A centrifugal pump consists of two basic parts, a rotating part and a stationary part. The rotating part includes the impeller and shaft, and the stationary part includes the casing, casing cover and bearing.

Exercise

1.（1）True.（2）True.（3）False.

2.（1）此外，离心泵通常不能输送含有蒸汽的液体。

（2）流体的动能从叶轮的中心到叶轮叶片的尖端增加。

（3）Piping arrangements when using centrifugal pumps are also usually simple.

Further Reading

Reading comprehension

1. No, intermittent discharge of fluid is obtained by a reciprocating pump.

2. No. Reciprocating and rotary pumps are self-priming pumps.

Part Ⅳ-Unit 10-Lesson Two Answers

Reading comprehension

1. Filtration may be viewed as an operation in which a heterogeneous mixture of a fluid and solid particles are separated by a filter medium that permits the flow of the fluid but retains the particles of the solid.

2. Plate-and-frame filters consist of porous plates that are held rigidly together in a frame. Feed slurry is pumped to the unit under pressure and flows in the press and into the bottom-corner duct of the frame. This duct has outlets into each of the frames, so the slurry fills the frames in parallel. The plates and frames are assembled alternately with filter cloths over each side of each plate. The assembly is held together as a unit by mechanical force applied hydraulically or by a screw.

3. Centrifugal filtration is used when the solids are easy to filter and a filter cake of low moisture content is desired.

Exercise

1.（1）True.（2）False. Rotary vacuum filters are used where a continuous operation is desirable.（3）True.

2.（1）The per unit of filtering surface of plate-and-frame filter is probably the cheapest of all the various filtration equipment.

（2）随着过滤的继续，滤饼会在介质的表面形成，这些颗粒被认为是在孔隙上架桥。

（3）制药行业也在类似的应用中使用压滤机来浓缩工艺浆液中的产品。

Further Reading

Reading comprehension

1. Microfiltration (MF) is a process that is used to filter very fine particles (smaller than several microns) in a suspension by using a membrane with pores that are smaller than the particles.

2. This separation method offers following advantages: It does not depend on any density difference between the cells and the media. The closed systems used are free from aerosol formation. There is a high retention of cells (>99.9%). There is no need for any filter aid.

Part Ⅳ-Unit 11-Lesson One Answers

Reading comprehension

1. A heat exchanger is a heat-transfer device that is used for the transfer of internal thermal energy between two or more fluids available at different temperatures.

2. To increase heat-transfer area, secondary surfaces known as fins may be attached to the primary surface.

3. Tubular heat exchangers—double pipe, shell and tube, coiled tube; Plate heat exchangers—gasketed, spiral, plate coil, lamella; Extended surface heat exchangers; Regenerators.

4. They are the first choice because of well-established procedures for design and manufacture from a wide variety of materials, many years of satisfactory service, and availability of codes and standards for design and fabrication.

5. The coiled tube heat exchanger is not cheap because of the material costs, high labor input in winding the tubes, and the central mandrel, which is not useful for heat transfer but increases the shell diameter.

Exercise

1.

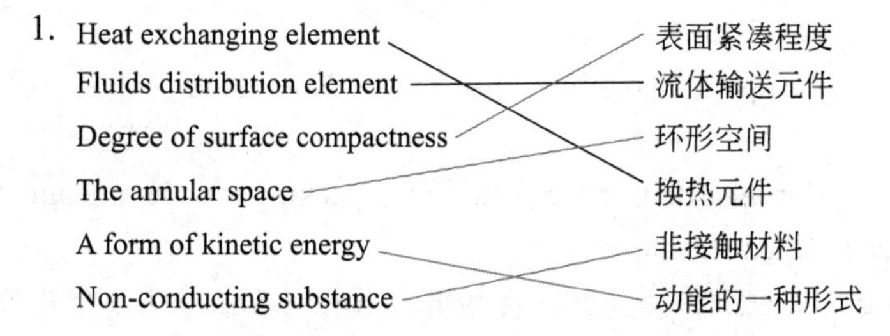

2.（1）材料的热导率差异很大，金属最大，非金属其次，液体更小，气体最小。

（2）通过辐射传递能量的独特之处在于不像传导和对流一样导热物质是必需的。

（3）The convective heat transfer coefficient when the fluid has a phase change is far greater than the convective heat transfer coefficient when there is no phase change.

（4）The high temperature liquid goes through the red pipe, and the low temperature liquid goes in the space between the blue and green pipes.

Further Reading

Reading comprehension

1. Heat, a form of kinetic energy, is transferred in three ways: (i) Conduction; (ii) Convection; (iii) Radiation.

2. Conduction.

3. The net energy that is transferred by radiation is equal to the difference between the radiations emitted and those absorbed.

Part Ⅳ-Unit 12-Lesson One Answers

Reading comprehension

1. Absorption is defined as the process of transfer of molecules of one substance directly to another substance. The process of absorption conventionally refers to the intimate contacting of a mixture of gases with a liquid so that part of one or more of the constituents of the gas will dissolve in the liquid.

2. An operating line is obtained by a mass balance around the column.

3. The rich solution enters the stripping unit and the volatile solute is stripped from solution by either reducing the pressure, increasing the temperature, using a stripping gas to remove the vapor solute dissolved in the solvent, or any combination of these process changes.

4. Packed columns are usually vertical columns that have been filled with packing or material of large surface area.

5. The liquid is distributed over and trickles down through the packed bed, thus exposing a large surface area to contact the gas.

Exercise

1.（1）True.（2）False.（3）True.

2.（1）In gas absorption operations, the choice of a particular solvent is also important.

（2）During installation prior to pouring the packing into the column, the column may first be filled with water.

（3）再分布器使液体离开壁并将其引向塔的中心以重新分布并与下一个较低部分接触。

Further Reading

Reading comprehension

1. Adsorption is a mass transfer process in which a solute is removed from a gas stream because it adheres to the surface of a solid. Absorption refers to the intimate contacting of a mixture of gases with a liquid so that part of one or more of the constituents of the gas will dissolve in the liquid.

2. Desorption is usually carried out by physical methods such as temperature increase, pressure drop, or exchange against another medium.

Part Ⅳ-Unit 12-Lesson Two Answers

Reading comprehension

1. The degree of difference in relative volatility of the components dictates the extent of separation of the mixture.

2. Batch distillation is typically chosen when it is not possible to run a continuous process due to limiting process constraints, the need to distill other process streams, or because the low frequency use of distillation does not warrant a unit devoted solely to a specific product or operation.

3. No, the reflux ratio should gradually be increased to ensure constant purity of the overhead product with time.

4. If the process requirement is to strip a volatile component from a relatively nonvolatile solvent, the rectifying (bottom) section may be omitted, and the unit is then called a stripping column.

5. In a partial condenser, the reflux will be in equilibrium with the vapor leaving the condenser, and is considered to be an equilibrium stage in the development of the operating line when estimating the column height.

Exercise

1.（1）Distillation: Distillation may be defined as the separation of the components of a liquid feed mixture by a process involving partial vaporization through the application of heat.

（2）Batch distillation: Batch distillation is generally carried out in the upward operating mode, in which the initial mixture is heated in the distillation vessel and the individual components are collected overhead one after the other in order of their volatilities, starting with the lowest boiling fraction.

（3）Continuous distillation: In continuous distillation, a feed mixture is introduced to a column where vapor rising up the column is contacted with liquid flowing downward (which is provided by condensing the vapor at the top of the column). This process removes or absorbs the less volatile (heavier) components from the vapor, thus effectively enriching the vapor with the more volatile (lighter) components.

（4）Reflux ratio：The ratio of the amount of reflux returned to the column to the distillate product collected is known as the reflux ratio.

2.（1）相对挥发度可以描述为特定化合物相对于另一种化合物蒸发度的比率。

（2）The greater the relative volatilities, the easier the separation.

Further Reading

Reading comprehension

1. The liquid enters at the top and flows downward *via* gravity. On the way, it flows across each plate and through a downspout to the plate below. The gas passes upward through openings of one sort or another in the plate, then bubbles through the liquid to form a froth, disengages from the froth, and passes on to the next plate above. The overall effect is a multiple, countercurrent contact of gas and liquid.

2. Both excessive entrainment and weeping greatly influence tray efficiency and negatively impact overall column performance.

Part Ⅳ-Unit 12-Lesson Three Answers

Reading comprehension

1. Liquid extraction is usually selected when distillation or stripping is impractical or too costly; this situation occurs when the relative volatility for two components falls between 1.0 and 1.2.

2. The degree of separation that arises because of the aforementioned solubility difference of the solute in the two phases may be obtained by providing multiple stage countercurrent contacting and subsequent separation of the phases, similar to a distillation operation.

3. Countercurrent operation with a given amount of solvent has the greatest separation enhancement.

4. If the separation achieved is inadequate, it can be increased by either changing the phase ratio or by the addition of more contacting stages.

Exercise

1.（1）Liquid-liquid extraction (or liquid extraction) is a process for separating a solute from a solution employing the concentration driving force between two immiscible (non-dissolving) liquid phases.

（2）The solvent-lean, residual feed solution, with one or more constituents removed by extraction, is referred to as the raffinate.

（3）The solvent-rich solution containing the extracted solute(s) is the extract.

2.（1）根据应用的不同，液-液萃取工艺可以在更高的选择性或更低的溶剂用量和更低的能耗方面具有优势。

（2）There are several principles that can be used as a guide when choosing a solvent for a liquid extraction process.

（3）Liquid-liquid extraction may also be carried out in a batch operation.

Further Reading

Reading comprehension

1. The first way is to set up a means for recycling the solvent, which is already quite a common practice in preparative chromatography. The problems associated with the consumption of solvent can also guide the selection process in advance. A third approach aiming to reduce the environmental impact of solvents is the use of solvents from "green" chemistry and to minimize their consumption by increasing, for example, the extraction using microwave heating, ultrasound heating, and so on.

2. Extraction by supercritical CO_2 has the advantage of resulting in a spontaneous separation of substances extracted by simple depressurization.

Part Ⅳ-Unit 12-Lesson Four Answers

Reading comprehension

1. The drying of solids to remove moisture involves the simultaneous processes of heat and mass transfer. Heat is transferred from the gas to the solid (and liquid) in order to evaporate the liquid contained in the solid.

2. Surface liquid; Capillary-bound liquid; Dissolved or chemically bound.

3. No. After drying has proceeded for some time, the surface film begins to disappear and the rate of drying decreases.

4. The air drying rate of materials depends on various process parameters, of which the most important are temperature, airflow and velocity of drying air, relative humidity, initial and final air and product temperature, and material size.

5. Tunnel dryers; Rotary dryers; Spray dryers.

Exercise

1.（1）False.（2）False.（3）True.

2.（1）并流干燥导致高干燥速率并限制了最高产品温度，而逆流干燥允许实现较低的最终水分含量。

（2）It should be noted that drying is one method of separating a liquid from a solid.

Further Reading

Reading comprehension

1. Additional steps such as compaction and granulation must be used.

2. The choice of co- or countercurrent drying is determined by the product properties. Cocurrent drying leads to high drying rates and limits the maximum product temperatures, while countercurrent drying allows lower final moisture contents to be achieved.

Part Ⅴ-Unit 13-Lesson One Answers

Reading comprehension

1. The "triple bottom line" of Green Chemistry is sustainability in economic, social, and environmental performance.

2. The 12 well-known principles involve the various aspects of the synthetic chemistry and the chemical processes, such as the greenization of raw material, process and products; the production cost; the consumption of energy; and the safety techniques and so on.

Exercise

（1）绿色化学是减少或消除有害物质的使用或产生的化学产品和工艺的设计。

（2）An ounce of prevention is worth a pound of cure.

Further Reading

Reading comprehension

1. Catalysts are substances that accelerate reactions, typically by enabling chemical bonds to be broken and/or formed without being consumed in the process.

2. Catalysts are a tremendously important and active area of both fundamental and applied research.

Part Ⅴ-Unit 13-Lesson Two Answers

Reading comprehension

1. The total number of positive charge and negative charge on the numerical is always equal, so in the macro the charge is neutral and the plasma is neutral.

2. We usually express the charged particle energy in the plasma as electron temperature, the unit is eV.

3. Electromagnetic radiation is transmitted in waves or particles at different wavelengths and frequencies.

Exercise

1.（1）True.（2）True.（3）False.

2.（1）与传统的焚烧法相比，它不会产生具有强毒性和二次污染物的二噁英，并具有处理速度快、效果好、能耗低等特点。

（2）The spectrum is generally divided into seven regions in order of decreasing wavelength and increasing energy and frequency.

Further Reading

Reading comprehension

1. An ultrasonic pulse is generated in a particular direction. If there is an object in the path of this pulse, part or all of the pulse will be reflected back to the transmitter as an echo and can be detected through the receiver path. By measuring the difference in time between the pulse being transmitted and the echo

being received, it is possible to determine the distance.

2. It can be performed with centimeters to meters accuracy.

Part Ⅴ-Unit 13-Lesson Three Answers

Reading comprehension

1. Supercritical carbon dioxide (sCO_2) is a fluid state of carbon dioxide where it is held at or above its critical temperature and critical pressure. It likes a gas but with a density like that of a liquid.

2. Its main advantages are that it is non-toxic, non-flammable, and simpler separation from the starting materials.

Exercise

1.(1)False.(2)True.(3)True.

2.(1)它在化学工业中具有广泛的应用，还显著减少了能耗，并且可以减少许多关键工艺的温室气体排放。

(2)简而言之，必须考虑溶剂从摇篮到坟墓(或摇篮到摇篮，如果回收利用)的整个使用寿命的影响。

Further Reading

Reading comprehension

1. While ordinary liquids such as water and gasoline are predominantly made of electrically neutral molecules, ionic liquids are largely made of ions.

2. Ionic liquids have many potential applications, like electrolytes and sealants.

Part Ⅴ-Unit 13-Lesson Four Answers

Reading comprehension

1. Chemical safety includes all those policies, procedures and practices designed to minimize the risk of exposure to potentially hazardous chemicals.

2. In order to manage risk, a duty holder must do the following steps: identifying hazards, assessing risks, controlling risks, monitoring and review, emergency preparedness.

3. The identity of chemicals in the workplace can usually be determined by looking at the label and the safety data sheets (SDS), and reading what ingredients are in each chemical or product.

Exercise

1.(1)True.(2)False.(3)True.

2.(1)当与其他化学品混合、加热或处理不当时，许多化学品的危险性可能会增加。

(2)The type of risk assessment that should be conducted will depend on the nature of the work being performed.

Further Reading

Reading comprehension

1. PPE is personal protective equipment. Safety goggles, standard gloves, laboratory coats.

2. The Globally Harmonized System of Classification and Labelling of Chemicals are now used throughout much of the world.

参考文献

[1] CLAYDEN J, GREEVES N, WARREN S. Organic Chemistry[M]. 2nd ed. UK: Oxford University Press, 2012.

[2] HOUSECROFT C，SHARPE A. Inorganic Chemistry[M]. 5th ed. England: Pearson, 2018.

[3] ATKINS P. PAULA J, KEELER J. Physical Chemistry[M]. 11th ed. UK: Oxford University Press, 2018.

[4] 胡彩玲 . 物理化学 [M]. 北京：化学工业出版社，2017.

[5] DE VOE H. Thermodynamics and Chemistry[M]. 2nd ed. New York: Prentice-Hall, 2010.

[6] KENNETH S. Brief introduction to the three law of thermodynamics [J]. Journal of Chemical Education, 1975, 52(5):330-331.

[7] HIRSCH H. Rubber bands, free energy, and Le Châtelier's principle[J]. Journal of Chemical Education, 2002, 79(2):200A-200B.

[8] 陈善英 . 科技英语的文体特征实例分析 [J]. 江西金融职工大学学报，2006, 19: 173-175.

[9] WELTY J, RORRER D, FOSTER D. Fundamentals of Momentum, Heat and Mass Transfer[M]. 6th ed. New York: Wiley, 2013.

[10] MATLACK A S. Introduction to Green Chemsitry[M]. 2nd ed. Boca Raton: CRC Press, 2010.

[11] MULHOLLAND M. Applied Process Control[M]. Weinheim: Wiley-VCH, 2016.

[12] PROCHASKA C, THEODORE L. Introduction to Mathematical Methods for Environmental Engineers and Scientists[M]. USA: Scrivener Publishing, 2018.

[13] AHMED J, RAHMAN M S. Handbook of Food Process Design[M]. New Jersey: Wiley-Blackwell, 2012.

[14] VOGEL C H. Process Development[M]. Weinheim:Wiley-VCH, 2005.

[15] THEODORE L, DUPONT R, GANESAN K. Unit Operations in Environmental Engineering[M]. USA: Scrivener Publishing, 2017.

[16] 马俊波，张伟，阮红缨 . 乐学英语写作教程 [M]. 北京：外语教学与研究出版社，2019.
[17] 张道真 . 张道真英语语法大全 [M]. 北京：世界图书出版公司，2019.
[18] 樊才云，钟含春 . 科技术语翻译例析 [J]. 中国翻译，2003,(01):59-61.
[19] 韦孟芬 . 英语科技术语的词汇特征及翻译 [J]. 中国科技翻译，2014，27(01):5-7, 23.
[20] 余高峰 . 科技英语长句翻译技巧探析 [J]. 中国科技翻译，2012，25(03):1-3.